Agriculture Chemical User`s Manual

Guidelines and Priciples for resposible Agricultural Chemical Use

Edited and Updated by DPI & F:

Geoff Cowles

Jodi Crome

Rowan Lambourne

Paul Simson

Rick Webster

Consultants:

Nicholas Woods

The Centre for Pesticides Application and Safety
School of Agriculture and Horticulture
The University of Queensland, Gatton Campus
Gatton QLD 4343
Australia

United Book Prints

Reprinted in India by:

UNITED BOOK PRINTS
(Imprint of Scientific Publishers, INDIA)
4806/24, Ansari Road, Daryaganj,
New Delhi 110 002, INDIA
E-mail: info@scientificpub.com
Website : www.scientificpubonline.com

Print, 2021

This Publication Provides Guidlines and Principles for Resposible Agricultural Chemical Use

ISBN:9789383692224

ISBN:
eISBN:

9 789383 692224

Printed in India

Contents

LIST OF FIGURES

LIST OF TABLES

INTRODUCING THE AGRICULTURAL CHEMICAL USER'S MANUAL

Agricultural chemicals have an important role in the efficient and economic production of wholesome food and fibre products, reducing soil erosion and maintaining human health and lifestyles for an ever-increasing world population. Agricultural. chemicals are also known as pesticides and include herbicides insecticides and fungicides.

Imagine a world where whole crops are totally destroyed by fungi, bacteria, locusts or mice, or their yields severely reduced due o weeds which starve the crops of vital water and nutrients. It has been estimated that food supplies could immediately fall to 30% if the ravages of pests were unchecked. It is also hard to envisage a world where weeds or ants are allowed to overtake our recreational areas such as parks or sporting venues. These scenarios reinforce the need for adequate control measures to be available, including agricultural chemicals.

However, responsible chemical use is the key to ensuring that agricultural chemicals are beneficial to the community as a whole, and not just beneficial to users only. The community must have confidence that human health and the environment are protected when agricultural chemicals are used.

The Agricultural chemical users' manual has been published joy the Department of Primary Industries and Fisheries (DPI&F) as a resource about responsible agricultural chemical use. This manual is a tool for anyone involved in pest or weed control activities by providing users with sound scientific advice on a range of subjects including basic pest and weed identification, effective pest and weed control strategies, appropriate application technology, as well as ways to reduce the impacts of agricultural chemical spray drift. Drift of chemicals can be the source of considerable damage to crops and can affect: the marketability of livestock.

The manual has application as DPI&F"s approved study text for those seeking qualifications for a commercial operator's licence to apply ground applications of herbicides. In addition, the manual is a useful resource tool for farmers and growers wishing to improve their skills in using agricultural chemicals. Training providers who deliver agricultural chemical application courses (e.g. ChemCert, TAFE and other registered training organisations) will also find this manual beneficial.

Above all, the manual promotes responsible agricultural chemical use by all users.

It should be noted that this manual does not replace or surpass any State, Territory or Commonwealth law, and agricultural chemical users should always make themselves aware of and adhere to relevant laws.

CONTROLLING AGRICULTURAL PESTS

Pest control and pest management

Pests are living organisms that can cause injury or damage to crops, livestock and stored food. They can significantly reduce yields, increase the cost of production and, at worst, cause the complete failure of entire cropping systems. Not all living organisms are pests. Most organisms contribute to our cropping systems in beneficial ways. For example, bees are valuable as they transfer pollen for many plant species, and certain bacteria are essential for fixing nitrogen in soil and plants.

There are many ways of controlling pests. The aim is to minimise damage and keep pest populations below the level at which economic damage occurs. It is not always necessary to use agricultural chemicals. Some of the main pest control methods used in agriculture, forestry and public health industries are listed below.

Physical and mechanical control

Methods include using fire, hand-picking or mechanical cultivation for weed control, and using light traps to attract insects.

Cultural or managerial control

These are defined as techniques that manipulate cropping systems to provide good growing conditions. These techniques can also make the crops unattractive or unsuitable for pests to flourish. Cultural controls include managing farm hygiene (e.g. cleaning equipment between paddocks), using trap or companion crops to concentrate pest species, and crop rotation.

Biological control

This method of pest control uses living organisms (or products derived from them) to minimise the activity of a pest species, usually through parasitism, predation or disruption of normal metabolic activities. For example, Cactoblastis cactorum, a moth, was successfully introduced to control the prickly pear cactus; and toxins produced by the bacterium Bacillus thuringiensis are used widely to control certain caterpillars and mosquitoes.

Quarantine and exclusion

Physical or legal barriers can be used to prevent the movement of pests from 'infected' to 'non-infected' areas. For example, the movement of sugarcane from one production area to another is strictly regulated.

Genetic control

Plants can be bred or modified with pest resistant characteristics using traditional plant breeding techniques or more advanced genetic: engineering technologies. For example, plants can be bred that produce toxins effective against insect pests. Mechanical defensive structures such as hairy leaves and waxy cuticles help plants to resist insect attack. Plants can be bred that are resistant to common fungal pathogens.

Control of pests using agricultural chemicals

Chemical control is also an important method of pest control. There is a wide range of agricultural chemical products registered in Australia for the control of approximately 2200 insects, weeds and fungal pathogens.

Throughout this manual, the term 'agricultural chemical' is used. It is synonymous with the term 'pesticide'. Agricultural chemicals include insecticides, herbicides, fungicides and rodenticides.

Agricultural chemicals offer a number of control advantages, namely:

- Products registered for specific: pest or crop situations can be selected.
- Agricultural chemicals are often fast-acting and produce rapid results.
- When a pest species reaches an economic threshold, most agricultural chemicals can be applied rapidly and efficiently, often over large areas.

- Many agricultural chemicals are selective and enable specific organisms to be targeted and controlled.

However, agricultural chemicals should always be used with care. The convenience of agricultural chemicals has contributed in part to overuse and misuse in many parts of the world.

Care should be taken to prevent:

- resistance in target pests
- environmental damage through the movement of products in the air, soil and water
- chronic and acute health affects to the user
- reduction in numbers of beneficial organisms.

Users of agricultural chemicals have the responsibility of ensuring the correct product is applied, at the correct dose, at the right time and with minimum contamination of the environment. These factors are explained fully in this manual.

Spray application: The process

The job of a spray applicator is sometimes incorrectly simplified as the following:

(1) Take agricultural chemical. (2) Place in sprayer. (3) Spray desired target. (4) Hope pest dies.

After studying this manual, the reader will appreciate that the correct application of agricultural chemicals requires more thought, care and understanding than the above sentence might suggest.

Before the use of an agricultural chemical is considered, the correct target must be identified. The following important questions need to be asked:

- What is the intended target?
- What control or management practices pan be implemented?
- If an agricultural chemical is selected, which product should be used?
- How does the product work?
- Where should the agricultural chemical De applied?
- When should the agricultural chemical be applied?

Targets for agricultural chemicals may be identified as:

- soil

- plants or any parts of plants such as foliage, flowers, fruits or seeds
- actual pests such as insects, fungi or weeds.

Other factors that need to be considered are:

- the site of the target. This includes, for example, its position within a crop canopy, or, in the case of a fungus or insect, the position of an identified pest on a leaf or on the ground.
- the behaviour of the target. A weed might be regarded as a static pest. However, a successful insecticide application often requires a confident understanding of pest mobility, feeding habits and life cycle. For example, coastal salt-marsh mosquitoes are usually best controlled by placing larvicides into salt-water pools where the immature larvae can be controlled before pupation occurs.

Importantly, the spray applicator has to be confident that the product applied is reaching its intended target at the correct dosage and with minimum impact on the surrounding environment. In particular, the product must be applied in such a way that there will be good coverage of the target with minimal impact ors off-target and adjacent areas. This theme will be explored in this manual.

PEST BIOLOGY

WEEDS

Introduction

To ensure agricultural chemicals are used correctly, the target pest needs to be correctly identified. Users of agricultural chemicals also need to understand pest biology to recognise when control measures are required and when they are best implemented.

The term 'pest' includes weeds, insects and mites, and organisms that cause plant disease, such as fungi. Chemical users need to know how to identify pests and recognise the damage they can cause in agricultural situations. A good understanding of pest biology and different pest management strategies helps the user to choose the most appropriate strategy foil a particular pest problem.

What is a weed?

The most commonly recognised definition of a weed is: 'A plant growing where it is not wanted'.

Weeds have the ability to:

- compete with crops, pastures, recreational areas or forestry
- change natural ecosystems
- harbour pests or provide an alternative host for diseases
- cause harm by way of toxic or physical threat to humans and livestock
- induce taints in livestock products such as milk
- induce allergies
- create safety problems
- block waterways and water storage bodies
- increase the management cost of public and privately owned amenities

- interfere with services (telephones, power lines etc.)
- devalue land.

Most weeds in Australia came from other parts of the globe and were imported, either deliberately for commercial reason, or by mistake with other material. So, local biological control agents for the management or control may not exist.

WEED LIFE CYCLES

Short life cycle — less than two years

Weeds with short life cycles germinate from seed, produce a shoot and roots, flower, set seed and die. They survive from one generation to the next as seeds. Preventing them from seeding is a top control strategy.

There are three main short life weed types:

- **Ephemerals** germinate, grow, flower and shed seed in a few weeks or months, then die. They usually go through several generations in a year.
- **Annuals** complete their life cycle in a single season, either summer or winter. Their seeds wait until the same season the following year before germinating.
- **Biennials** have a two-year life span. In the first year, they usually grow in the shape of a low, flat rosette of leaves building up reserves in a taproot. In the second year, they produce a shoot and set seed before they die.

Weeds with a short life cycle usually produce large numbers of seeds that give rise to a large number of individual plants. Plants exposed to a non-lethal dose of herbicide are more likely to develop resistance.

Long life cycle — longer than two years

Perennials are weeds that go on living and producing seeds every year if seasonal conditions allow. They have a variety of survival methods to cope with climatic extremes.

They can be divided into two groups:

- **Soft or herbaceous perennials.** These go through a normal growth, flowering and seeding cycle but usually die back at the end of the growing season. They build up reserves that support regrowth from dormant buds when suitable growing conditions return.

- **Woody perennials.** These are weeds that produce a woody stem to support their leaves. The stem framework grows each year and is connected to a root system designed to provide a stable anchor for the increasing weight above ground.

Some perennials can spread without flowering; by having parts that can grow into identical individual plants:

- **Stolons** are horizontal stems that run above the ground. They also strike roots and produce stems at intervals.
- **Rhizomes** are horizontal stems that run underground. They spread out from the parent plant and produce adventitious roots and aboveground stems at intervals along their length (see Figure 2.1).

If stolons or rhizomes are cut as a result of cultivation or mechanical damage, a new individual plant will develop that is genetically identical in every way to the original parent.

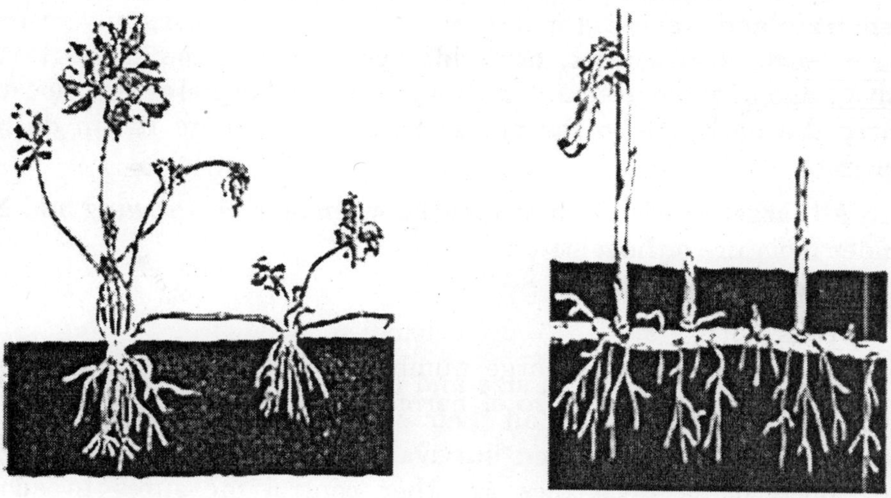

Figure 2.1 : Examples of stolons (left) and rhizomes (right)

Tubers are short, fleshy underground stems. They usually carry both food reserves and dormant buds and act as the survival form of the plant between growing seasons (see Figure 2.2).

Corms and bulbs are arrangements of stems tightly surrounded by specialised leaves. Corms have a thick stem and scale-like leaves, whereas bulbs have a normal-sized stem surrounded by very fleshy leaves. Chemical management of such plants is difficult because it is not easy for agricultural chemicals to penetrate them (see Figure 2.2).

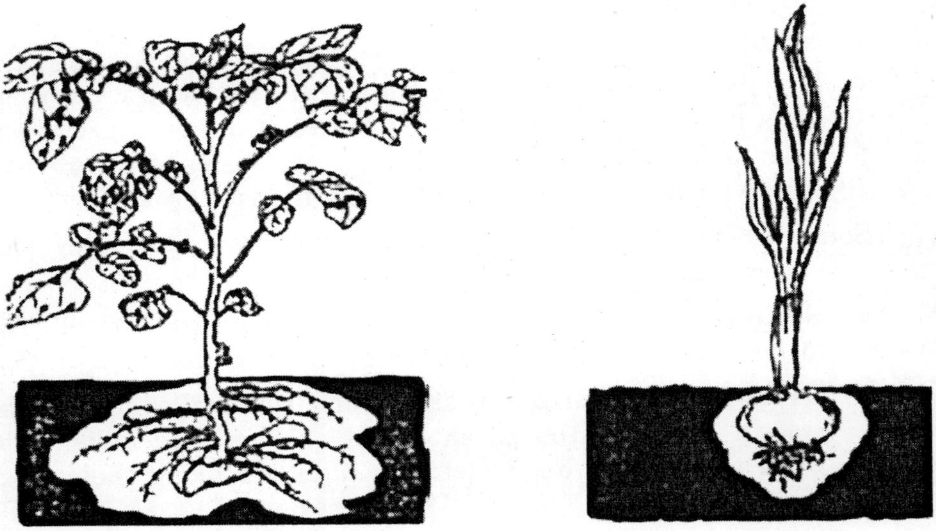

Figure 2.2 : Examples of tubers (left) and corms (right)

For a herbicide to be effective against perennials, it must penetrate and travel through the whole plant including any reserves. To achieve good control, herbicide sprays should be timed to take advantage of sap movement within the plant that helps translocate or carry the herbicide to the parts of the plant where it can do most harm.

All target weeds are best sprayed when actively growing and NOT under moisture or heat stress.

Weed botany

Weeds vary greatly in size and complexity. The control of weeds depends on knowing how all their parts, including roots, stems and leaves, support growth and survival. This enables strategies to be developed using herbicides or other control measures to achieve effective results. Weed life cycles and growth habits influence the selection and application of control strategies. Having a basic understanding of plant structure and the function of different parts of a plant helps the user understand how a herbicide reaches where it does its damage.

Weeds have different growth habits but like all plants, they have green leaves and the ability to extract energy from sunlight. Green leaves contain chlorophyll. The chemical process that chlorophyll uses to produce sugars is called photosynthesis.

Plant groups

Plant groups can be divided into two main growing types, the grasses (monocotyledons or monocots) and the broad-leaved plants (dicotyledons or dicots).

Their main distinguishing features are listed below:

Monocots

Monocots are plants that have a single first seed leaf and include grasses and sedges. Other features of monocots are:

- narrow leaves with parallel veins
- growing points close to the ground.

Dicots

Dicots are plants that have two first seed leaves. The shape of these seed leaves is a key feature when identifying a weed species.

Other features of dicots are:

- branched veins in mature leaves
- sap-transporting organs nearer to the outside of the stem than monocots. With woody dicot plants, this feature enables control by ring-barking and basal stem spraying.

Basic plant structure

Most weeds have:

- roots to anchor them in the ground and to draw nutrients from the surrounding soil
- a stem or framework of stems to support leaves
- flowers to produce pollen, seeds or both.

Transport organs help move sap within the plant. Moisture and dissolved soil nutrients move upwards and outwards in the xylem (see Figure 2.3). Sugars and carbohydrates manufactured in the leaves move upwards and downwards in the phloem.

If a herbicide is intended to damage some growth system of a weed, it has to enter the plant via the leaves or the roots and then use these transport routes to reach the site where it causes damage.

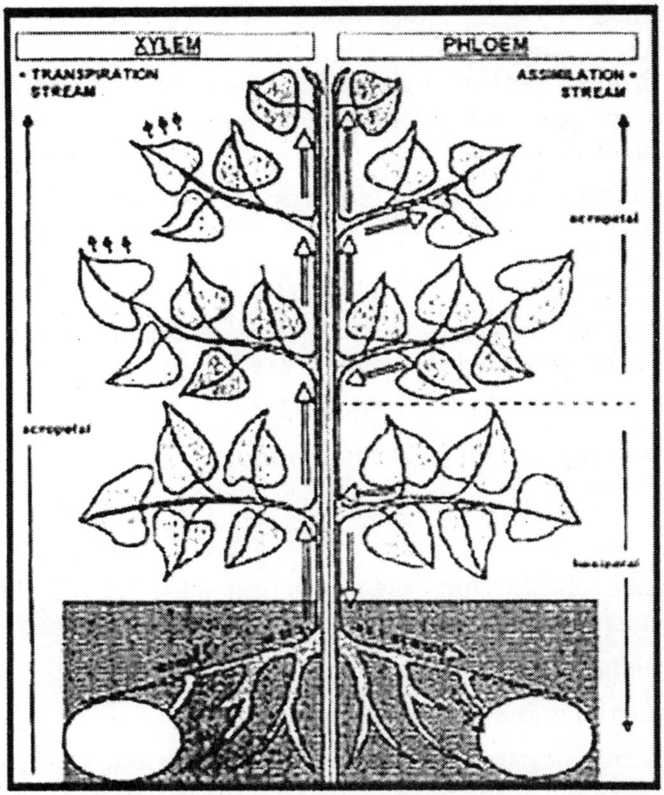

Figure 2.3 Basic plant structure and vascular system

The root system

The shape of root systems varies. However, dicots generally develop a central anchoring taproot with side branches. Monocot roots are less formally structured and are usually described as being fibrous. Adventitious roots-grasses with underground runners and roots appearing at joints along the runner-become important if the runner is cut by cultivation as they can be utilised to support new individual plants.

The stem

The function of the stem is to support and orientate the leaves to the light (to enable photosynthesis) as well as to provide enough mechanical strength to keep the plant upright.

The stem encloses the transport organs and may provide a platform for the flowers. Stems make pollination (by insects or wind) easier and seed dispersal more efficient.

In woody weeds, a protective layer of bark usually covers the stem. This is dead material and thickens as the weed gets older. In soft herbaceous plants, the stem is often green because it contains chlorophyll in the outer cell walls.

Transport systems

Sap movement occurs by means of two distinct and different 'plumbing' systems that run parallel with each other. One runs upwards and outwards from the roots to the leaves and is called the xylem. It consists of dead woody tubes. Its function is to transport solutions that penetrate the tube. It also gives some support strength to the stem. Soil-applied herbicides use the xylem to spread through their targets.

The transport system used for moving the sugars and other products of photosynthesis from the leaves downwards and around the plant is called the phloem. Unlike the xylem, phloem consists of living cells. This means that agricultural chemicals applied to leaves have to pass through living cell walls to travel (translocate) in the plant. Cell walls act as barriers to some chemicals.

The xylem and the phloem together are called the vascular system.

The leaves

The main function of the leaf is to provide housing and protection for the green chlorophyll containing cells that carry out photosynthesis. The leaf surface is usually protected by a waxy layer called the cuticle. The cuticle has a number of pores scattered over it called stomata. The stomata allow gas exchange to occur between the plant tissues and the surrounding atmosphere. These pores are normally open, in order to retain moisture, but may close when the humidity levels are low.

The weed as a herbicide target

It is always useful to know when weed seeds are germinating so that appropriate early control can be taken.

Plants are dynamic, complex organisms that depend for their health and growth on a number of biochemical processes. Herbicides are designed to target one or more biochemical processes. Spraying aims to deliver the right dose of herbicide when the chances of getting the active molecule to a specific target are greatest. If a herbicide does not translocate easily, the addition of a wetting agent or surfactant to the spray solution can help.

A successful result depends on delivering the herbicide spray under the right weather conditions. Weeds under stress, especially drought stress, will not respond satisfactorily because growth processes are slowed to a minimum until moisture is available. Under these conditions, herbicides have difficulty in penetrating the target weed, moving inside it or finding an active biological receptor.

Spraying stressed weeds results in poor control.

Insects and mites

Insects and their arachnid relatives (spiders, mites and ticks) are the most numerous and diverse group of animals on earth. Most do not compete directly with humans for food and fibre. Some even play a useful role and these species are referred to as beneficials. Pest insects, however, can do harm in a number of ways:

Pest example	**Plant damage**
Caterpillars, grasshoppers and locusts	Damage leaves by chewing
Citrus leaf miner	Mine inside leaf tissue
Codling moth in apples, sugarcane stem borers	Bore into fruit and stems
Aphids, thrips	Suck out plant sap
Fruit flies	Lay eggs in developing fruit

The type of damage caused by insects depends on the structure of the insect's mouthparts and its mechanism of feeding. Caterpillars, for example, have hard-biting mouthparts capable of tearing and chewing leaves.

Other insects have a 'beak' designed to puncture the leaf cuticle and suck the plant sap from the phloem, e.g. aphids. Any viruses in sap taken up by sucking pests can be transmitted to the next plant on which it feeds.

Characteristics of insects and mites

Both insects and mites are in the grouping known as arthropods. These animals are characterised by having bodies divided into segments and a hard outer covering or exoskeleton. They have flexible joints and paired limbs. Insects and mites can be distinguished in the following ways.

Adult insects have:

- three body segments (head, thorax and abdomen)
- six legs
- one pair of antennae
- may or may not have wings.

Adult mites have:

- two body segments (cephalothorax and abdomen)
- eight legs (usually)
- no antennae
- no wings.

The biology of insect and mite pests

Insects are great opportunists. They combine mobility, which allows them to be in the right place at the right time, with the capacity to increase their numbers rapidly to take advantage of favourable conditions. As well as this, they have a hard outer coating (exoskeleton) which provides good protection against the elements (and agricultural chemicals).

The insect life cycle involves distinct changes in body appearance and behaviour patterns.

There are three types of insect life cycle:

- The **complete** metamorphosis life cycle includes larval, pupa) and adult stages and is typical of many important pest species such as *Helicoverpa* spp. (see Figure 2.4). The egg hatches into a larva or caterpillar. The larva grows through several stages of increasing size known as instars. At each of these stages, the old skin is shed (a process called moulting) and a new covering develops to protect the larger body.

  Eventually, the larva is ready for the dramatic change into the adult form. This change or metamorphosis takes place inside a body stage called the pupa (chrysalis). Inside the pupa, the larva alters shape into the adult stage. When the transformation is complete, the adult emerges.

- In the **no metamorphosis** life cycle, the insect hatches from the egg as a small version of the adult and progresses through a nuriber of increasing size changes to adulthood while maintaining a similar appearance.
- In the **incomplete metamorphosis** life cycle, the nymph that hatches from the egg is similar in shape and appearance to the adult, but lacks some adult features such as wings, which develop as it matures through moult stages. Locusts are examples of such insects.

The prime role of adults is to reproduce. Often mobility provided by wings enables mates to be found and allows travel to new potential feeding grounds for the next generation. Adults usually place their eggs in situations that give the hatchlings the best chance of survival.

The time taken to complete the life cycle varies with species and the season.

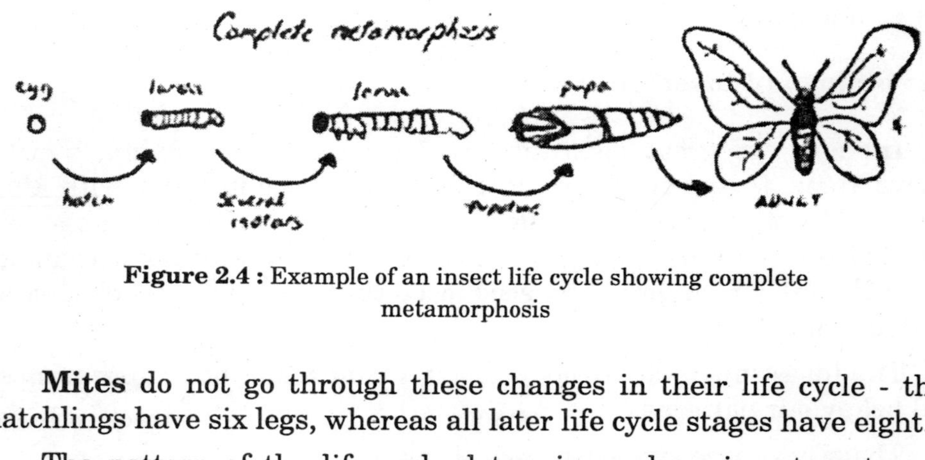

Figure 2.4 : Example of an insect life cycle showing complete metamorphosis

Mites do not go through these changes in their life cycle - the hatchlings have six legs, whereas all later life cycle stages have eight.

The pattern of the life cycle determines where insect pests are most vulnerable to agricultural chemical control.

INSECT TARGET STRUCTURES

Eggs

Eggs are laid singly or in clusters and their shape and size can assist in identifying species. Eggs are hard for agricultural chemical sprays to penetrate and can be inaccessible (e.g. in leaf rolls or in buds). Observing egg colour change is an important way of timing insecticide sprays because it indicates when a larval mass emergence can be expected.

Larvae

The early use of an insecticide, a type of agricultural chemical intended to control insects, can reduce damage caused by larvae. In general, tie smaller the larva, the less chemical needed for a lethal dose.

The decision to spray means that the larvae should be present in large enough numbers to cause potential economic damage and any beneficial species present will be incapable of reducing those numbers before serious damage has occurred.

Pupae

The pupa, like the egg, is also difficult for agricultural chemical sprays to penetrate and can be a difficult target because it usually occurs in protected situations (e.g. leaf rolls or in the soil).

Adults

The adult can be an elusive target when agricultural chemical sprays are applied due to mobility. Insecticides are effective where a deposit landing on a plant is stable enough for it to be eaten or crawled on by the insect pest.

Beneficial insects

Some insect species attack insect pests and therefore are useful in integrated crop protection strategies. Wasps in particular are a group containing several important beneficial families that attack other insects.

Beneficial insects have the following features:

- Different species parasitise different insect pest targets, including aphids, eggs, larvae (caterpillars), pupae of moths and butterflies, whiteflies and scale insects.
- They lay their eggs on the pest target that provides enough food for them to complete their own development.
- They fly to their targets, making them less vulnerable to chemically sprayed surfaces.
- If their population is large enough and the parasites have beneficial attributes, they can significantly reduce pest numbers. However, they build up their numbers at a slower

rate than the pests they are attacking, so they need supplementary help to reduce the pests to acceptable levels.

Consequently, if insecticide sprays are selected that do as little harm as possible to the beneficial species, they can be used to amplify the impact of a particular insecticide.

This can be achieved in two ways:

- Insecticides should be applied that are known to have little effect on beneficial species.
- If practical, some unsprayed areas should be maintained near production areas so that these areas can act as a refuge and reservoir for both the beneficials and their prey. If a sudden increase in pest pressure requires the application of a non-selective insecticide, these reservoirs can help restore beneficial numbers rapidly, by providing a source from a viable population.

If the deliberate release of beneficial species is part of the management plan, allowance should be made for any residues from previous sprays to break down, so that the introduced beneficial insects do not die also.

Small organisms that cause plant disease

Plant diseases are caused by small organisms such as fungi, bacteria, viruses and nematodes. For plant disease to become established, the correct combination of host, environment and organism must be present. This relationship is often referred to as the disease triangle.

Before fungicides can be used in the management of plant disease, an understanding of the disease triangle is essential. Often good hygiene and the use of crop rotations are more effective and economic than chemical control.

Fungi

Plant diseases caused by fungi can damage crop yield and quality during crop growth and after harvest.

Fungi have the following features:

- They survive by attacking living plants and dead plant material.
- They consist of a simple vegetative body called a mycelium made up of branched filaments called hyphae. These hyphae

may or may not have internal separating walls that are called septa.

- Fungi contain no chlorophyll and cannot make their own food, so they are totally dependent on other sources (e.g. a host) for their energy requirements.
- Most reproduce by means of spores that may be sexual or non-sexual. The incredible variety of shapes and sizes of these spores and how they are produced form the basis of the classification of species.
- Sexually reproducing fungi sometimes overcome host resistance or become resistant to fungicides through sexual recombination.
- Non-sexual spores spread the disease dining the plant's growing season. Many fungi survive from one season to the next in the form of a sexual spore.
- The location for survival between seasons varies and may be:
  - on or in the host plant
  - on or in plant debris left after harvest
  - in the soil
  - in the host plant's seed.

The location of the spores usually determines the type of fungicidal treatment needed for the control of diseases. Examples of fungicide application treatment include leaf sprays, fungicidal (fruit or root) dips, soil fumigation and seed treatment.

Preventing fungal damage

All parts of a plant may exhibit fungal diseases. However, individual species are often limited in the plant parts they attack. Often diseases are named for the symptoms they cause. Leaf blights and rusts, stem cankers, fruit scabs, diebacks and freckle all give an idea of what is happening to the parts of the plant under attack.

Managing plant protection to prevent disease can involve many different strategies. Only some of these strategies include agricultural chemical application.

The main methods of managing fungal diseases are:

- making sure growing conditions do not favour soil-borne disease by improving drainage and avoiding disease-infested areas

- rotating between crops that do not host the same diseases, to break the cycle of soil-borne diseases
- the use of disease-resistant crop varieties
- the use of disease-free planting material
- the strategic use of fungicides when required, including soil fumigation to ensure a clean seedbed
- the use of strict sanitation procedures by destroying all plant material at the end of each growing season
- disinfecting all equipment after use in crops and between blocks.

REFERENCES

Swain, J. & Reed, G. (1998) Principles of Pest Management, Agsafe Limited, Canberra.

Hartmann, H.T. *et al.* (1998) Plant science: Growth, development, and Utilization of Cultivated Plants. 2nd ed., Prentice-Hall, New Jersey.

Kerruish, R.M. & Unger, P.W. (1991) Plant Protection 1: Pests, Diseases and Weeds, Root Rot Press, Canberra.

Low, T. (1999) Feral Future, Viking, Victoria.

Parry, D.W. (1990) Plant Pathology in Agriculture, Cambridge University Press, Cambridge.

Wells, C.H. (1986) Weed Control Notes, Queensland Department of Forestry, Brisbane.

AGRICULTURAL CHEMICAL CLASSIFICATION

Introduction

Agricultural chemicals (sometimes called agchems, agrochemicals, pesticides or farm chemicals) are substances used for controlling or managing pests in specific crops and situations.

Agricultural chemicals are biologically active substances with a potential for harm as well as good. Their use requires responsible management, application and handling, taking into consideration their impacts on people, property and the environment.

Part of responsible management when using agricultural chemicals is having a basic understanding of how these products are classified, so that the right product is used correctly.

Agricultural chemicals are normally classified according to their use and the nature of their intended target. Examples include herbicides, insecticides and fungicides. These classifications can be sub-divided into types to describe when particular chemicals are applied and according to the way they work (mode of action). This method of classification helps form the basis of resistance management strategies as recommended by the agricultural and veterinary (agvet) chemical industry body, Avcare. Further information on these strategies and the classification systems are available in the appendices.

Herbicides

Herbicides are designed to kill or control unwanted plants (or weeds) in crops, pasture or non-crop situations such as roadsides or recreational areas, e.g. parklands. The many herbicides registered for use in Australia can be classified according to their recognised modes of action (see Appendix A) and application timing. The following are examples:

- **Pre-emergence** herbicides are applied before crop or weed emergence.
- **Post-emergence** herbicides are applied after crop or weed emergence.
- **Contact** herbicides kill plant tissue where they make contact with the target, e.g. paraquat (found in Gramoxone) and diquat (found in Reglone).
- **Translocated** (Systemic) herbicides are able to move within the target weed. They can move from the contact site to another site within the weed via the weed's vascular system. Movement may start from droplets, containing herbicide, contacting the leaves (downward movement from the leaves), or the roots may take the herbicide, in solution, from the soil (upward movement from the roots). Glyphosate (found in Roundup) is an example of a herbicide that works through downward movement from the leaves. Atrazine (found in Gesaprim) is an example of a herbicide that works through upward movement from the roots.

Insecticides

Insecticides are designed to kill insects and arachnids (spiders, ticks and mites). Many insecticides work by interfering with the insect's nervous system in some way.

Individual insects or arachnids, in a group, may become resistant to an insecticide through natural selection-the insecticide will no longer kill them. These insects or arachnids that become resistant to a particular insecticide may also be resistant to any other insecticides of the same group. Avoiding continued use of insecticides from the same group will help prevent insect resistance.

Insecticide types are based on their mode of action (see Appendix D), the nature of the killing agent, the name of the specific target species or the life cycle stage of the target.

The main classifications of insecticides are:

Classification by mode of action

- **Stomach poisons.** These insecticides land on a leaf surface and stay long enough to be eaten by the target larvae, e.g. endosulfan (Thiodan).

- **Contact poisons.** These insecticides need to thoroughly cover the target so that physical contact between the pest species and the insecticide car occur. Good coverage is usually achieved by the application of a large number of (small) droplets, e.g. methomyl (Lannate).
- **Lures.** These insecticides are usually formulated as a bait. The active ingredient is usually mixed with an attractant to lure the insect to a lethal dose of insecticide or to a sticky surface, e.g. fruit fly baits.
- **Insect growth regulators (IGRs).** These insecticides interfere with the way insects moult between instars, or they prevent the smooth transition from a pupa to an adult, e.g. methoprene (Altosid).
- **Microbial insecticides.** These insecticides are usually non-chemical bio-agents that have been formulated and packaged in a convenient way so that they can be handled and applied exactly like regular chemical insecticides. Members of this group include *Bacillus thuringiensis* (Dipel) and viruses such as *Helicoverpa zea nucleopolyhedrovirus* (Gemstar).
- **Transgenic insecticidal crops.** A transgenic crop is one into which genetic material from an unrelated species has been deliberately introduced by genetic engineering, e.g. Boilgard that is used on cotton. This insecticide contains the gene from *Bacillus thoringiensis*, which acts as a toxin in the stomachs of insect pests such as *Helicoverpa armigera* and *Helicoverpa puntigera*.

Classification by target species

- Aphicides kill aphids.
- Acaricides kill ticks.
- Miticides kill mites.
- Ovicides kill insect and mite ova (eggs).

Fungicides

Fungicides are agricultural chemicals used to protect a crop against attack from fungal pathogens. There are two main types of fungicides:

- **Protestant.** These fungicides are deposited as a covering film on the leaf or plant tissue. They act like a protective coat and remain on the plant until a fungal spore lands on the leaf, germinates and acquires a dose of the agricultural chemical as it tries to penetrate the leaf, e.g. mancozeb (Dithane).

- **Systemic.** These fungicides are able to move within the plant. Some systemic fungicides move in the plant's vascular tissue, e.g. propiconazole (Tilt).

Others

The following are other agricultural chemical groups based on their intended target:

- **Rodenticides** control rodents such as rats and mice.
- **Molluscicides** control molluscs such as slugs and snails.
- **Insect and bird repellents** repel insects and birds.
- **Bactericides** are antiseptics that attack and suppress bacterial growth, e.g. dairy tank cleansers and cane knife sterilisers.
- **Nematicides** control nematodes such as root-knot nematode.

PRODUCT FORMULATIONS

Introduction

Agricultural chemicals are generally made up of several components which together make up the chemical formulation. The product is sold to users in this form. The most important part of any formulation is the active ingredient (also referred to as the active constituent), which is the part that is biologically active on target pests. Other components of the formulation, called the inert ingredients, are designed to make the active ingredient more efficient and to keep the product stable until it is required for use.

Types of formulation

There are a variety of product formulations. The hazards associated with their use, risks to the environment, effectiveness in pest control, and cost depend in part on the formulation type.

When an agricultural chemical product is being manufactured, the properties of the active ingredient will usually dictate the choice of formulation that can be used to produce a stable, consistent and marketable product.

Common formulations Solutions

These are true liquids that contain the active ingredient dissolved in either water (water-based aqueous concentrates) or a solvent that mixes with water. They contain no suspended particles.

Solutions may be concentrated and need further dilution before application or they may be packaged in a convenient form ready for immediate spraying.

The advantages and disadvantage; of solutions depend on the solvents used, the concentration of the active ingredient and the type of application equipment used. Many of the advantages and disadvantages associated with emulsifiable concentrates (ECs), listed below, apply also to solutions.

Ultra low volume (ULV)

ULV formulations are a special type of solution that can be applied straight from the container with no further dilution. They consist of the active ingredient dissolved in organic solvents and oil. These products can be measured out at low volumes and are often applied using aircraft.

Advantages:

- Aircraft can spray a very large area without refueling, as spray volume application rates are very low.
- Spraying equipment needs to be refilled less frequently, as ULVs are applied with little or no dilution.
- Evaporation rate of the agricultural chemical is reduced because, when ULVs are diluted, oil rather than water is used.

Disadvantages:

- They require equipment specifically designed to apply small quantities of agricultural chemical over a large area.
- Extreme care must be taken when calibrating the specialised equipment because of the frequently high concentration of active ingredient.

Emulsifiable concentrates (ECs)

ECs are another type of liquid that contain an emulsifier in their formulations to enable the active ingredient to dissolve in an organic solvent and then mix evenly in water when a spray mixture is prepared. The spray mixture is an emulsion and is always milky white in colour.

Advantages:

- They are relatively easy to measure, handle, transport and store.
- Once mixed, they need little agitation.

- They are not abrasive and do not block application equipment; however, some solvents cause wear of washers and seals.

Disadvantages:

- The solvent system used may be toxic to plants at high temperatures.
- The solvent may increase the risk of skin irritation and absorption if accidental exposure occurs to the agricultural chemical user or livestock.
- Some solvents are highly flammable and result in the product being classified as a dangerous good.
- There may be disposal problems with some containers, in particular, formulations containing petroleum oil.

Soluble powders (SPs)

Soluble powder formulations dissolve in water to form a true solution.

Advantages:

- They are easy to store and transport.
- They pose a lower risk to non-target plants than ECs.
- They can be packed in disposable packages.

Disadvantages:

- Some finely divided powders can be hazardous and care should be taken to avoid breathing in the powder when measuring oat doses.
- Measuring is difficult unless pre-packs are used.

Wettable powders (WPs)

WPs are designed to disperse in water to form a suspension that is then applied as a spray. The advantage of this formulation is outweighed by a number of disadvantages, listed below, which have led to a decline in the popularity of WPs.

Advantages:

- They are a convenient way of fucking stable high concentrations (up to 80%) of the active ingredient.

Disadvantages:

- Constant agitation in the spray tank is needed to avoid uneven dosing caused by particle settlement.
- The suspended particles are abrasive and produce accelerated wear in nozzles and pumps.
- Nozzles and filters can be blocked by suspended solids.
- Many WPs require careful pre-mixing with a little water to ensure even dispersion and this process can be difficult with some alkaline bore waters.
- Measuring out by weight can bi› hazardous unless pre-packs are used and then the pre-packs have to coincide with the dose required.

Suspension concentrates or flowable concentrates (SCs)

SCs were introduced to overcome some of the handling problems caused by WPs.

With this formulation, the active ingredient is milled to a finer size than WPs. It is then packed off as a suspension in water to be further diluted to make up a spray. Special care is needed to make sure the suspension does not settle in storage. Packs of SCs must always be shaken vigorously before measuring out doses.

Advantages:

- They are abrasive but, because of their finer particle size, there is less chance of nozzle or filter blockages occurring.
- They can be dispensed accurately by volume in a measuring jug.
- They do not have the loose dust problems associated with WPs.

Disadvantages:

- Less of the active material can be incorporated in the formulation without causing stability problems.

Water dispersible granules (WGs)

This type of formulation has become popular with agricultural chemical manufacturers, as new technologies have produced micro-granules that can carry high concentrations of active ingredients; this is acceptable with low toxicity products.

- WGs avoid the problems of dust generation.
- Fine milling of the base ingredients of the formulation, prior to actual granule formation, minimises blockages.

Granules (GRs)

Free flowing granular formulations have been used for a long time as a means of applying agricultural chemicals to control soil-borne pests or to apply agricultural chemicals without water in remote areas that depend on rainfall for later activation.

Some granules are formulated using a polymer matrix that degrades at a predictable rate releasing small doses of active ingredient over an extended period. These are known as slow release granules and they were introduced to extend the control period of agricultural chemicals.

Granules provide a relatively safe means of handling hazardous agricultural chemicals as the formulation involves scattering a small amount of active ingredient through a much larger bulk of inert material, usually clay.

Advantages:

- They are ready to use without mixing and are easy to apply.
- Application does not involve carrying water, therefore reducing soil compression and permitting application in hard-to-access areas.
- They usually have little or no dust associated with them and therefore present a lower risk to agricultural chemical users than WPs.
- The application equipment needed to disperse them is relatively cheap when compared with hydraulic sprayers.
- Granules can penetrate foliage to reach the soil surface more easily than spray droplets.

Disadvantages:

- They are more expensive as the amount of active ingredient incorporated is at a lower level per unit weight.
- They may require soil incorporation or follow-up rain before becoming active.
- They may present a hazard to non-target species, especially birds.

OTHER FORMULATIONS

Aerosol dispensers (AEs)

Aerosols provide convenience, but at a higher price. They are common in domestic situations for applying and dispensing insecticides. With aerosol dispensers it is difficult to control where the spray fall-out will land. Also small droplets are more likely to be inhaled. The formulations often contain a flammable propellant under pressure that poses potential hazards if the container is punctured or incinerated.

Fumigants

These hazardous formulations are slowly being phased out. They can penetrate target areas very efficiently and usually only involve a single application. However, fumigants are extremely hazardous to use. Most require special training in safe handling and also require the operator to wear carefully selected personal protective equipment (PPE). Confining fumigating gases to the desired area of action can sometimes be difficult.

Water soluble crystals

This formulation is particularly useful for 2,4-D herbicide products. The crystals are packed in water-soluble plastic pre-packs. These are dropped into the spray vat to dissolve and form the spray. The agricultural chemical concentrate is not exposed to the handler at any time.

Emulsions (oil in water) (EWs)

FAO and WHO define this formula ion as a stable emulsion of active ingredient(s) in an aqueous phase, intended for dilution with water before use. Some agricultural chemicals for home garden use are sold in this form.

Capsule suspensions (CSs)

These are a newer formulation where the active ingredient is contained in polymeric microcapsules dispersed as a stable suspension in a liquid such as water. They are normally intended for dilution with water before use. An advantage oil this formulation over other liquid formulations is that agricultural chemical user exposure to the toxic active ingredient is reduced during the mixing and application stages. Most microcapsules provide slow release of the active ingredient and help prolong the residual activity of the diluted product, and some burst as soon as the spray dries.

INERT INGREDIENTS

Adjuvants

An adjuvant is any chemical added to a formulation or spray mix for the purpose of improving its performance or stability. Adjuvants can act as an aid to identification (e.g. a dye added to seed treatment formulations to permit identification of treated seed).

The main types of adjuvants are listed below:

- **Synergists** are chemicals added to a formulation to enhance the performance of the active ingredient, although alone, they have little or no activity. Examples include:
    - o piperonyl butoxide added to some pyrethroids to improve the knock-down of flying insects
    - o ammonium thiocyanate added to amitrole products to enhance the herbicidal effect
    - o organosilicone penetrants (see also wetting agents) added to herbicide formulations.
- **Buffers** are chemicals that can aster the pH (acidity or alkalinity) of water used as a spray carrier from its normal pH. Some agricultural chemicals perform more consistently under slightly acidic conditions. Where the local water supply is alkaline, acidifying buffers will lower the pH. Commercially available spray buffers usually contain a surfactant (see also wetting agents).
- **Sequestering agents** are used in some formulations to overcome the problems caused by hard water. Hard water contains significant amounts of calcium and magnesium salts. In a number of phenoxy herbicide formulations (e.g. 2,4-D)

EDTA (ethylene diaminetetra acetic acid) is added to prevent reactions that would otherwise produce insoluble (solid) salts settling out of solution.

- **Humectants** are added to flowable formulations to prevent the suspension being deposited on the internal walls of flowable packs. If the suspension is deposited, it can dry out into hard flakes that resist re-dispersion and can eventually cause filter and nozzle blockage. Humectants attract moisture that may be present as vapour above the main bulk of the suspension concentrate within the packs. Deposits still occur from liquid movement in transport but remain as a deposit that can easily be re-mixed when the pack is shaken.
- **Anti-foaming agents or defoamers** can be useful in preventing excess foam production when spray tanks are agitated and when spray tanks are recharged with high-pressure water. Wetting agent traces left from the previous vat load are the cause of the foam being produced.

Wetting agents

Wetting agents include formulation and spray adjuvants and are also called wetters, surfactants (surface active agents), spreaders or stickers. They sometimes form part of a formulation and sometimes are added to the spray mix at application. Unless the label directions contain specific advice about this, the use of additional wetting agent is wasteful and may cause problems such as excessive naming in the spray tank and damage to non-target plants.

Wetting agents lower the surface tension of the agricultural chemical being applied so that the liquid spreads as a film over the plant surface rather than resting as individual droplets. In this way, better contact is made between the spray and the target. Also, there is a special group of wetters and surfactants known as organosilicone penetrants that aid the penetration of the active ingredient through the leaf surface to enhance weed and pest control. An example is the addition of Pulse penetrant to glyphosate sprays to improve control of hard-to-kill weeds such as bracken fern.

There are three basic types of wetting agents that are distinguished by their chemistry: nonionic, cationic and anionic. Most wetting agents currently available on the market for use with agricultural chemical products are non-ionic and are the most appropriate to use in the majority of situations. The only exception is where it may be necessary to alter or stabilise the pH of the spray mixture, e.g. by using an acidifying buffer surfactant where the local water supply is alkaline. Many industrial surfactants and household detergents are either cationic or anionic and should never be used with agricultural chemicals as they may cause stability problems in the spray mix and damage non-target plants.

Examples of typical formulations

Examples of typical agricultural chemical formulations are listed below:

Emulsifiable concentrate (Insecticide)

Dimethoate technical (96% active)	417 g/L (providing 400 g/L dimethoate)
Organic solvent	589 g/L
Emulsifying agent	50 g/L

Suspension concentrate (Herbicide)

Diuron technical (94%)	532 g/L (providing 500 g/L diuron)
Dispersant	80 g/L
Humectant	110 g/L
Defoamer	5 g/L
Viscosity modifier	5 g/L
Water	300 g/L

Dispersible granules (Herbicide)

Chlorsulfuron technical (95%)	790 g/kg (providing 750 g/kg chlorsulfuron)
Naphthalene sulphonates	100 g/kg
Kaolin	110 g/kg

REFERENCES

drumMuster (2001) drumMuster Collection manual, Version 2, July, Canberra.

Barrett, M. *et al.* (2002) Agsafe accreditation training manual, Version 9, Agsafe Limited, Canberra.

FAO/WHO Joint Meeting on Pesticide Specifications (JMPS) (2002) Manual on development and use of FAO & WHO specifications for pesticides: Pesticide specifications, 1st ed., Rome.

PRODUCT LABELS

INTRODUCTION

The label format for agricultural chemicals is largely determined by legislation. The product label's main function is to explain how a product can be used most effectively to remove a pest problem. The label also provides enough information to make sure the product is used and disposed of in a safe and efficient manner. **A label is a legally binding document, approved by the Australian Pesticides and Veterinary Medicines Authority (APVMA), formerly the National Registration Authority (NRA).**

A label has a number of key sections. Below is an example of a typical agricultural chemical product label. The label's various segments are identified. A description of each of the identified segments is also given. It has been prepared by the APVMA and is reproduced from their document, Ag labelling code (code of practice for labeling agricultural chemicals).

WHAT A PRODUCT LABEL CONTAINS

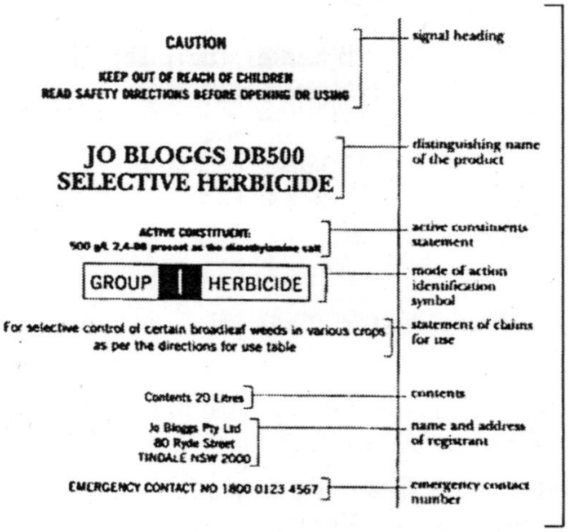

Figure. 5.1 Example of a typical product label

DIRECTIONS FOR USE

directions for use

RESTRAINTS

DO NOT apply when rain is expected within 4 hours.

DO NOT apply to crops or weeds which are stressed by drought or cold, frosty conditions.

restraints

CROP	WEEDS	RATE L/ha	CRITICAL COMMENTS
Barley, Oats, Wheat undersown with Lucerne, Medics or Subterranean clover	Refer to weeds tables 1 and 2 following	2.1–3.2	Apply when crop is from tillering after the main shoot has 5 leaves until before the boot stage; and when lucerne, clover and medics have 1–8 trifoliate leaves (1–5 trifoliate leaves in Vic and Tas). Apply when weeds are at the 2–6 leaf stage, use lower rate for 2–4 leaf weeds under good growing conditions.
Lucerne seedling and established	Refer to weeds table 1 following	1–2	Apply when seedling lucerne has 1–8 trifoliate leaves (1–5 trifoliate leaves in Vic and Tas); cut or graze established lucerne before application to avoid leaf distortion, and apply when crop regrowth is 10–13 cm high. Apply when weeds are at the 2–6 leaf stage.
	Refer to weeds table 2 following	2.1–3.2	Apply as above. Use lower rate when weeds are at 2–4 leaf stage under good growing conditions.

use table

NOT TO BE USED FOR ANY PURPOSE, OR IN ANY MANNER, CONTRARY TO THIS LABEL UNLESS AUTHORISED UNDER APPROPRIATE LEGISLATION

'not to be used' statement

WITHHOLDING PERIOD

DO NOT GRAZE OR CUT FOR STOCK FOOD FOR 7 DAYS AFTER APPLICATION

withholding period

Figure. 5.1 Example of a typical product label (continued)

weeds table

WEEDS TABLES

Note: The rate tolerated by the crop/pasture limits the range of weeds which may be controlled.

WEEDS TABLE 1

WEEDS CONTROLLED	RATE L/ha	CRITICAL COMMENTS
Bathurst Burr	1-3.2	Spray young seedlings only
Californian Burr	1-3.2	Spray young seedlings only
Docks	1-3.2	Spray up to rosette stage
Fat Hen	1	Spray up to rosette stage.
Noogoora Burr	1	Spray young seedlings only

WEEDS TABLE 2

WEEDS CONTROLLED	RATE L/ha	CRITICAL COMMENTS
Capeweed	3.2	Spray very young seedlings only
Fumitory	3.2	Spray young seedlings only
Paterson's Curse	2.1-3.2	Spray young seedlings only
Slender Thistle	2.1-3.2	Spray up to rosette stage

general instructions

GENERAL INSTRUCTIONS

resistant weeds warning

Resistant Weeds Warning

Jo Bloggs DB500 is a member of the phenoxy group of herbicides. Its mode of action is to disrupt the plant cell growth. For weed resistance management, Jo Bloggs DB500 is a Group I herbicide.

Some naturally occurring weed biotypes resistant to this product and other disruptors of plant cell growth may exist through normal genetic variability in any weed population. The resistant individuals can eventually dominate the weed population if these herbicides are used repeatedly. These resistant weeds will not be controlled by this product or other disruptor of plant cell growth herbicides.

Since the occurrence of resistant weeds is difficult to detect prior to use, Jo Bloggs Pty Limited accepts no liability for any losses that may result from the failure of this product to control resistant weeds.

general instructions on mixing and application

Mixing

Mix concentrate with the required volume of water in the spray tank. DO NOT use hard water. Use mains or rain water for mixing. If unavailable, use bore water with a hardness of less than 1,000 mg/L (expressed as calcium carbonate equivalent). If the hardness exceeds this level, an additive may be used to soften the water.

Application

Boom Sprayer

Ensure sprayer is fitted with fan jets. Calibrate prior to use to deliver 110-220 L spray/ha. Check height to ensure even distribution. Use a minimum 110 L water/ha. Wash spray equipment thoroughly after use.

Aircraft

Use 10 L water/ha.

Figure. 5.1 Example of a typical product label (continued)

Compatibility

This product is compatible with diuron and dimethoate, although the effectiveness of dimethoate may be reduced.

When tank mixing, add the compatible product when nearly full, while agitating. Agitate vigorously if allowed to stand. Test all mixtures before using commercial quantities.

PROTECTION OF CROPS, NATIVE AND OTHER NON-TARGET PLANTS

DO NOT apply under weather conditions or from spraying equipment that may cause spray to drift onto nearby susceptible plants/crops, cropping lands or pastures. Avoid spray drift and vapour movement onto susceptible crops such as cotton, tobacco, tomatoes, vines, lupins, fruit trees and ornamentals.

PROTECTION OF WILDLIFE, FISH, CRUSTACEANS AND ENVIRONMENT

DO NOT contaminate streams, rivers or waterways with the chemical or used containers.

STORAGE AND DISPOSAL

Store in the closed, original container in a cool, well ventilated area. DO NOT store for prolonged periods in direct sunlight.

Triple or preferably pressure rinse containers before disposal. Add rinsings to spray tank. Do not dispose of undiluted chemicals on site. If recycling, replace cap and return clean containers to recycler or designated collection point.

If not recycling, break, crush or puncture and bury empty containers in a local authority landfill. If no landfill is available, bury the container below 500 mm in a disposal pit specifically marked and set up for this purpose, clear of waterways, desirable vegetation and tree roots. Empty containers and product should not be burnt.

SAFETY DIRECTIONS

Poisonous if swallowed. May irritate the eyes and skin. The product is alkaline. Avoid contact with eyes and skin. DO NOT inhale the spray mist. When preparing spray wear cotton overalls buttoned to the neck and wrist, washable hat, elbow length PVC gloves and face shield. When using the prepared spray wear cotton overalls buttoned to the neck and wrist and a washable hat. If product on skin, immediately wash area with soap and water. After use and before eating, drinking or smoking, wash hands, arms and face thoroughly with soap and water. After each day's use, wash gloves, face shield and contaminated clothing.

FIRST AID

If poisoning occurs, contact a doctor or Poisons Information Centre. If swallowed and more than 15 minutes from a hospital, induce vomiting, preferably using Ipecac Syrup APF.

MATERIAL SAFETY DATA SHEET

Additional information is listed in the Material Safety Data Sheet which can be obtained from the supplier.

Batch: DOM:

NRA Approval No: 49357/01

general instructions

general instructions on compatibility (2.9)

protection statements

disposal and storage statement

safety directions

first aid instructions

MSDS reference

batch number

date of manufacture

NRA approval number

Figure. 5.1 Example of a typical product label

The main panel

The signal heading

This appears prominently at the top of the main panel above the trade name. The signal heading shown is determined by the particular poison schedule, the active ingredient(s) or other components as classified under health legislation (See Table 5.1).

Table 5.1 : Signal heading statements and the toxicity level indicated (Source : Standard for the Uniform Scheduling of Drugs and Poisons (SUSDP) No. 18)

Signal heading statements	Poison schedule	General description
DANGEROUS POISON KEEP OUT OF THE REACH OF CHILDREN READ SAFETY DIRECTIONS BEFORE USE	7	Substance with a high potential for causing harm at low exposure and which require special precautions during manufacture, handling or use. They should be available only to specialised or authorised users who have the skills necessary to handle them safely. Special regulations restricting their availability, possession, storage or use may apply.
POISON KEEP OUT OF THE REACH OF CHILDREN READ SAFETY DIRECTIONS BEFORE USE	6	Substance with a moderate potential for causing harm. The potential for harm can be reduced through the use of distinctive packaging with strong warnings and safety directions on the label.
CAUTION KEEP OUT OF THE REACH OF THE CHILDREN READ SAFETY DIRECTIONS BEFORE USE	5	Substances with a low potential for causing harm. The potential for harm can be reduced through the use of appropriate packaging with simple warnings and safety directions on the label.
PRESCRIPTION ANIMAL REMEDY KEEP OUT OF THE REACH OF CHILDREN READ SAFETY DIRECTIONS BEFORE USE	4	Substances where the use or supply should be by or on the order of persons permitted by State or Territory legislation to prescribe, e.g. veterinary surgeons.

Notes

1. The size of the signal heading depends on the size of the label. It must be in a contrasting colour so that it is clearly visible and easy to read.
2. Under legislation, the words 'KEEP OUT OF THE REACH OF CHILDREN' mean that the product must be stored and displayed, in public access areas, at least 1.2 metres above the floor. It is essential that in all situations, extreme care must be taken to prevent children from having access to all agricultural chemicals.
3. For agricultural, domestic and industrial poisons (such as pesticides), schedules 5, 6 and 7 represent increasingly strict container and labeling requirements with special regulatory controls over the availability of schedule 7 poisons.
4. Many agricultural chemicals contain ingredients that are not included in any of the poison schedules. In most cases, the manufacturer elects to show signal wording such as KEEP OUT OF THE REACH OF CHILDREN and READ SAFETY DIRECTIONS BEFORE OPENING, although it may not be mandatory to do so.
5. Some hazardous products have precautionary statements immediately below the statement 'KEEP OUT OFTHE REACH OF CHILDREN'. An example is 'CAN KILL IF SWALLOWED'.

The distinguishing name or registered trade name

This name (consisting of brands, trademarks, numbers and descriptive terms), displayed on the main label, distinguishes a product from other similar products.

The active constituent

The active constituent is also known as the active ingredient. This is the common chemical name of the active part of the formulation. Its details are placed under the trade name. The concentration of the active constituent present, expressed in grams per kilogram (g/kg) or grams per litre (g/L), is follower by the active constituent's common name. If the active constituent is present in a particular form, then the form is also included, e.g. 'present as the dimethylamine salt'. If more than one active constituent is present, they are included together, along with details of any scheduled solvent in the formulation. If the active constituent is an organophosphate or carbamate insecticide, the active constituent statement will be followed by the words: 'An anticholinesterase compound'.

Mode of action identifying symbol

This symbol is a letter of the alphabet and/or number, indicating the relevant mode of action grouping of the agricultural chemical The symbol is placed below the active constituent statement, e.g. 'GROUP H HERBICIDE'. This symbol helps when selecting agricultural chemicals for pest-resistance management programs (see Section 7).

Statement of claims for use

The central panel carries a brief statement summarising the purpose of the product.

Restricted chemical products

Certain products, nominated by the APVMA, will have the following statement after the claims for use: 'Restricted chemical product - only to be supplied to or used by an authorised person'. Authorised persons are often required to hold specific qualifications.

Net contents

The volume or weight of the formulated product in the pack appears on the lower main panel.

Registrant contact details

Name, address and emergency contact number of the registrant (manufacturer, formulator or distributor).

Ancillary panels or leaflet contents

Directions for use

The directions for use panels include instructions on how, when and where the product is to be used.

It may also include limitations on use that are intended to minimise hazards to human health, the environment, crops and livestock. This label section may have several sub-sections.

Restraints

Restraints are statements on the label that set limitations for the use of a product. They are applicable to all approved situations, such as the time taken for a sprayed agricultural chemical to be absorbed (e.g. 'DO NOT apply if rain expected within four hours').

Table of directions

Most labels include a table with the following column headings:

Crop/ situation	Pest target	State registration status	Application rate	Critical comments

The table is always followed by the statement: 'NOT TO BE USED FOR ANY PURPOSE, OR IN ANY MANNER, CONTRARY TO THIS LABEL UNLESS AUTHORISED UNDER APPROPRIATE LEGISLATION'.

This statement serves as a warning to users of the agricultural chemical that the product must only be used in accordance with label instructions. To do otherwise could result in the user committing an offence in Queensland under the Chemical Usage (Agricultural and Veterinary) Control Act 1988. This Act controls how agricultural and veterinary chemical products can be used in Queensland.

Below is a brief outline of each component in the table of directions:

- The **crop/situation** column sets out the crop (e.g. wheat, sugarcane) or situation (e.g. commercial and industrial areas, rights of way, around agricultural buildings) to be treated by the agricultural chemical.
- The **pest target** column sets out the pest, disease or weed to be controlled in the crop or situation being treated.
- The **state registration status** column provides the name of the States and Territories where the product may legally be used in the crop or situation to control the particular pest target. For example, if 'Qld' (Queensland) is not listed but 'NSW' (New South Wales) is listed, the product cannot be used for the particular crop or situation to control the pest target in Queensland but can be used in New South Wales.
- The **application rate** column sets out the amount of the product to be used to control a particular pest in the crop or situation.
- The **critical comments** column contains important information about how the user can maximise the desirable

effects and minimise the undesirable effects of the product they are using. For example, it could include an instruction that the product be applied at the four-to-six leaf stage of a certain weed, because it is more effective at that stage of the weed's growth.

Other limitations

This label panel may include warnings such as the product being too hazardous for use in the home garden.

Withholding period

'The minimum recommended interval that must elapse between the last application of a chemical product to a crop, pasture or animal, and the harvesting, grazing, cutting or slaughtering thereof, or the collection of milk and eggs for human consumption or the collection of fibre (as the case may be). It is part of the directions for use within the concepts of good agricultural practice in the use of agricultural chemicals products.' (Ag labeling code, 2001). An example might read: 'DO NOT HARVEST FOR 7 DAYS AFTER APPLICATION'.

Re-entry period

This is the time after spraying when a person must not enter the crop or situation area that has been treated with an agricultural chemical without wearing appropriate protective clothing.

General instructions

These are use instructions that vary according to the management that is needed to get the best results. Details could be provided, for example, on volumes of water per hectare, compatibility, mixing instructions and preferred spray application equipment.

Precautionary statements intended to limit the risk to human health from approved uses, such as exposure to the product, may be included here, in addition to the label safety directions.

Protection statements

These are statements that aim to safeguard off-target sensitive crops and limit adverse impact on wildlife, fish and the environment through responsible spray drift management.

Storage and disposal

Storage instructions will vary depending on the product that is used. In the case of schedule 7 products, the labels require the following statement on storage: 'Store in a locked room or place away from children, animals, Food, feedstuffs, seed and fertilisers'.

Disposal instructions must be followed.

Safety directions

These directions normally have three sections:

1. The risks and symptoms associated with exposure to the product.
2. A list of what personal protective equipment (PPE) should be used when handling the product.
3. Personal hygiene procedures as well as what to do if accidental contact is made with the product, whether in the concentrated form or after dilution, or if the product is ingested or inhaled.

First aid

This label panel follows the safety directions. It sets out emergency procedures to be carried out if a person has been exposed to the agricultural chemical, either by contact, ingestion or inhalation.

Individual pack identification

Every label must carry three items of information that, together, can identify the contents:

1. The **batch number,** which indicates the particular manufacturing batch from which the package of the product was derived.
2. The **date of manufacture** or the **expiry date,** depending on the product. The date of manufacture indicates when the batch of the product was made. The expiry date indicates when the product should no longer be used.
3. The **APVMA approval number,** which is unique to the product and pack size.

Dangerous goods symbols and emergency information

If the labeled product is classified as a dangerous good, the label will carry the following information:

- **correct shipping name** of the product, as listed in the Australian Dangerous Goods Code (the ADG Code)
- **UN number** (United Nations number) allocated to the product
- appropriate **Hazchem Code**
- diamond-shaped **dangerous goods class label**
- **emergency contact information.**

REFERENCE

National Registration Authority for Agricultural and Veterinary Chemicals (Australia) (2001) Ag labelling code: code of practice on labelling agricultural chemical products, Canberra.

PRODUCT SAFETY, TRANSPORT, STORAGE AND DISPOSAL

Introduction

The Workplace Health and Safety Act 1995 assigns obligations on employers and self-employed persons to not only manage risk to their own workplace health and safety but also to the health and safety of others arising out of the conduct of their business or undertaking.

Product safety considerations

Hazards and risks

Hazards - A hazard is something that has the potential to cause harm.

Risks - A risk is the likelihood of a hazard actually causing harm. In the case of agricultural chemicals, the risk to health usual y increases with the severity of the hazard, the amount used and the duration and frequency of exposure.

Material safety data sheets (MSDSs)

These documents are required for any agricultural chemical defined as a hazardous substance or dangerous goods under relevant legislation (Workplace Health and Safety Regulation 1997 and Dangerous Good Safety Management Regulation 2001).

Legislation states that the manufacturer or importer must:

- prepare an MSDS containing key information to users on the management of risk and meeting safety obligations
- review the MSDS at least once every five years
- amend the MSDS whenever necessary to ensure it contains current information

- provide a copy of the MSDS to each person the material is supplied to. In addition, the supplier must, in turn, give a copy of the MSDS to the end user and other designated persons, on request.

For an MSDS to be relevant:

- it should be read and understood before use
- it must refer to the actual formulation involved in the emergency
- it must be current.

An MSDS may contain the following segments:

- **Identification section.** Lists the names and identification codes used by the manufacturer, the UN (United Nations) number, and the dangerous goods class and subsidiary risk codes. It also provides information about the physical and chemical nature of the product.
- **Health hazard information.** Describe poisoning symptoms and detailed first aid, including advice to the doctor.
- **Precautions for use.** Details exposure standards, engineering controls, flammability and appropriate personal protective equipment (PPE) to be worn when handling the product.
- **Safe handling information.** Provides storage and transport information, procedures for dealing with spills and disposal and fire or explosion hazard advice, including hazardous decomposition products.
- **Other information.** Covers such matters as ecotoxicity, biodegradability and persistence in soil or water.
- **Contact point.** Gives a contact point in case of emergency.
- **Date of issue.** Indicates the date of issue.

Risks associated with agricultural chemicals used 'on farm'

Risks to people handling these products mad depend on:

- toxicity
- formulation that may include a volatile solvent
- flammability
- corrosiveness
- concentration of the active constituent present

- pack size and design
- how the product is handled, especially dining the mixing process
- workplace situation where spray solutions are prepared
- field situation where application takes place
- precautions taken to prevent exposure both during application and re-entry into the treated area
- training and competency of those handling the products
- weather conditions at the time of the spraying
- condition of the spray equipment being used
- state of mind and health of the person when preparing the spraying or when spraying.

Those unrelated to the workplace must not be exposed to the agricultural chemical. The chemical's toxicity, the condition of the spray equipment and the weather conditions at the time of spraying can lead to the risk of exposure if spray drift occurs.

Chemical exposure routes

Exposure routes list how agriculture chemicals can enter the human body.

There are three routes by which the body may be exposed to agricultural chemicals:

1. Inhalation. Inhalation is the fastest way of transferring a volatile gas from the atmosphere to the blood stream. Inhalation is also the exposure route for dusts. Adult human lungs have a working surface area of about 70 square metres and can absorb vapors from agricultural chemicals.

2. Ingestion. Good personal hygiene is the way to avoid chemical residues on contaminated hands reaching the blood stream via hand-held food, drinking or smoking. It is illegal to place agricultural chemicals in rood or drink containers. These products must always be kept in original containers.

3. Skin contact and eye splash. Skin contact and eye splash are the most common methods of exposure to agricultural chemicals in workplaces. The speed of uptake is faster under warm conditions and slows when it is cold. The rate also varies with parts of the body.

Table 6.1. Rates of absorption relative to the forearm (rated 1)

Body Part	Rate of uptake	Body part	Rate of uptake
Forearm	1	Scalp	3.7
Palms of hand	1.3	Forehead	4.2
Upper foot	1.6	Ear canal	5.4
Abdomen	2.1	Genitals	11.8
		Eyes	12

Note

1. The rate of skin uptake increases with formulations containing petroleum-derived solvents.
2. Eye splash can result in topical damage as well as rapid entry to the blood stream.

Good strategies to reduce agricultural chemical exposure

- Select agricultural chemicals on the basis of low toxicity and short persistence.
- Design a dedicated mixing facility that minimises the risk of exposure to the user.
- Follow label and MSDS directions when selecting personal protective equipment (PPE).
- Follow label directions on re-entry periods, i.e. do not enter a sprayed crop without wearing appropriate PPE until this time has passed.
- Mix and apply according to label instructions.
- Clean up any accidental spills immediately.
- Keep careful records of the whole operation, including health records of those handling agricultural chemicals regularly.

Personal protective equipment (PPE)

The selection of PPE will be determined by label instructions and MSDS information. The following table provides a guide to the PPE an operator should consider when using agricultural chemicals (see Table 6.2):

Table 6.2 : Personal protective equipment (PPE)

PPE item	Comments
Washable cotton hat or Built-in overall hood	Head covering to prevent scalp and hair exposure
Goggles	To give complete eye protection
Face shield	To give protection against face splash
Respirator which may have activated carbon P1, P2 or P3 Filter systems	Half or full-face respirator incorporating a cartridge filter system with dust, particulate and organic vapour filter options
Powered air- purifying full helmet	Necessary for high-risk situations and for spray applicators with beards. Powered helmets provide filtered air under positive pressure to the user. Respirator cartridges should always be suitable for the agricultural chemicals that are to be used.
Overalls	Buttoned at the wrist and neck these must be clean at the start of each day, splash proof and worn outside the boots
Gloves	Chemical proof, preferably unlined and elbow length
Apron	A full-length plastic apron gives added frontal protection when mixing concentrates
Boots	Rubber or PVC preferably with steel toe caps

Users should use PPE items in accordance with the manufacturers' specifications.

Maintenance of personal protective equipment

- All PPE should be cleaned as soon as possible after use.
- Soaking overalls in a slightly alkaline solution (e.g. 30 mL household bleach in 4 L water) will prevent residues becoming 'fixed' in the fabric. Soaking will also break down any organophosphate or carbamate residues that may be present.
- A similar solution may be used to wipe over other items of PPE before storage.
- Routinely check all items after use for wear and tear so that replacements can be obtained before the equipment is next required.
- A respirator cartridge should always be stored in a clean airtight container to prevent the activated carbon filter losing

useful life by exposure to dust and other contaminants in the atmosphere.

- The usage time of respirators should be recorded and they should be tested regularly for efficiency. This can be done by placing a dab of acetone or perfume on the outer side of the cartridge, donning the respirator and observing whether any odour can be detected.

Grading agricultural chemicals

The APVMA assesses the reports provided by chemical manufacturers on the toxicity of the active constituents of agricultural chemical products as part of the manufacturer's (registrant's) overall registration application package.

The toxicity is measured in experimental animals. There is a relationship between a chemical dose and the size (biomass) of the animal. For a given dose, a larger animal will show greater tolerance than a smaller one, so doses are always expressed in milligrams per kilogram liveweight. The figure used to compare the toxicity of chemicals is called the LD50 (Lethal Dose 50%). This is the dose of active constituent, expressed in milligrams per kilogram liveweight, needed to cause the death of half the test animal groin (rats, unless otherwise stated).

Note

1. The LD_{50} is based on the active constituent and not the formulated product.
2. The rat being tested may vary ire its tolerance of the chemical, so direct extrapolations to humans are not justified beyond a good indication of toxicity risk level.
3. LD_{50} figures developed as doses consumed by animals are known as acute oral LD_{50s}. Those developed from skin exposure are known as dermal LD_{50s}.

Table 6.3 shows examples of the relative toxicity of a number of products as expressed by the rat oral LD_{50}. The majority of agricultural chemicals fail within the 500 to 5000 mg/kg range.

Table 6.3. Conventional toxicity grades

LD_{50} range	Acute oral LD_{50} mg/kg	Chemical name	Example	Poison schedule	Comments on toxicity and availability
>15000 low toxicity					A few biocontrol agents
5000 to 15000	13700	ethyl alcohol	whisky	NS *	Very low toxicity. A few chemical products
	5000	triflumuron	Zapp	5	
500 to 5000	4300	glyphosate	Roundup	5	Slightly toxic. The majority of chemical products are in this range
	3300	sodium chloride	common salt	NS *	
	1750	acetyl salicyclic acid	asprin	2	
	1700	maldison	Malathion	5 or 6	
	1440	copper oxychloride	Cuprox	5 or 6	
50 to 500	400	carbaryl	Bugmaster	5 or 6	Moderately toxic. A small number of products
	150	paraquat	Gramoxone	7	
5 to 50	35	famphur	Warbex	6 or 7	Toxic. A few products still marketed
	32	coumaphos	Asuntol	6 or 7	
	5	amphetamine	'Speed'	8	
<5 high toxicity	1	aldicarb	Temik	7	Very toxic

* NS : not scheduled

Acute and chronic poisoning

Acute poisoning involves sudden exposure to a large amount of chemical. The symptoms of acute poisoning follow rapidly after exposure, so, often, there is little doubt about the source and appropriate action can be taken.

Chronic poisoning can be equally dangerous but is less obvious. Symptoms may just be a vague feeling of being unwell. Exposure occurs due to numerous, small sub-lethal doses over a period of time. Common causes are inhalation of very small droplets or undetected skin contact during handling and application of chemicals in spraying. Some chemicals are cumulative and slowly build up over a number of different exposure events to damaging levels in the body.

Symptoms may include:

- fatigue
- headache
- blurred vision
- nausea
- sweating
- coma leading to death.

It is wise to seek medical advice if feeling unwell at any time after handling agricultural chemicals.

It is prudent to try to anticipate possible accident scenarios and to have an emergency plan in place ready for rapid response. Seek medical advice to help put together such a plan.

Agricultural chemical residues

Agricultural chemical sprays produce residues that persist for various periods. Factors influencing persistence include:

- chemistry of the active constituent
- dose applied
- environmental conditions prevailing.

It is illegal to market any sprayed pr:)duce with residues above the maximum residue limit (MRL).

'The maximum residue limit (MRL) is defined as the maximum concentration of a residue, resulting from the officially authorised safe use of an agricultural or veterinary chemical, that is recommended to be legally permitted or recognised as acceptable in or on a food, agricultural commodity [product], or animal feed' (MRL standard, NRA 1996)

Observe the withholding period to prevent residues in agricultural products going above the MRL. The dose needed to kill a pest can result in residues above the MRL occurring in the produce immediately after application. A withholding period is, therefore, needed for the chemical residue level to decrease before crops are harvested, animals slaughtered or milk and eggs collected (See definition of withholding period in Section 5).

Transporting agricultural chemicals

The following provisions are mandatory when transporting agricultural chemicals by road. For on-farm transport, the provisions are strongly recommended but are not mandatory. Some agricultural chemicals are classified as dangerous goods and are subject to the Australian dangerous goods code.

- Full compliance with the Code may not be a legal obligation until the loads exceed one tonne.
- Compliance involves appropriate documentation, defined responsibilities and appropriate placarding (printed markings of vehicle).
- If a vehicle is used regularly to transport agricultural chemicals. It should contain an emergency kit consisting of appropriate FPE, reflector hazard warning signs, a suitable fire extinguisher, first-aid materials and a spill kit consisting of shovel, broom and a bag of neutralising agent such as hydrated lime (Agsafe accreditation training manual, 2002).

Those loading and unloading agricultural chemicals should heed the following advice:

Loading

- Never transport agricultural chemicals in the same cabin space as people, pets and human food. Remember: 'Ute it. Don't boot it'.
- Always check the agricultural chemical containers for corrosion and leaks.
- Check that the agricultural chemical containers have complete labels.
- Distribute the load evenly and secure it to prevent movement.
- Secure agricultural chemicals in vehicles so that they don't roll around.

Unloading

- Check that the load is complete.
- If there are spillages, clean them up immediately.

Storing agricultural chemicals

Wherever possible, obtain agricultural chemicals when they are needed rather than store them long-term for future use.

When planning the storage of agricultural chemicals, consider the following:

- **Location.** The store should be free of flood threat and be located at least 15 metres from the property boundary and to metres from other dwellings. Vegetation should be kept clear for at least three metres around the building.
- **Construction.** The store should be made of fire-resistant materials and have a sealed floor, which should be bunded to contain any leakage or spillage of stored product. Good ventilation prevents build-up of toxic or flammable fumes. Most stores should contain shelving to keep stock off the concrete floor. Bunding is defined as the physical retention of fire-fighting water or spillage. Sometimes the bunded area can be designed so that it drains to a pit where spills and contaminated water can be treated and made harmless. Additional retention volume cap sometimes be provided by bunding the loading and unloading area and other paved ground outside the premises.
- **Services.** A reliable water supply to the store is both a safety feature to service an emergency washing shower or pose, and also a convenience to fill spray vats if the mixing area is nearby.
- **Security.** The store should always be kept locked when not in use and clearly placarded (labelled or marked with important information about the agricultural chemicals being stored there).
- **Emergency equipment.** Arrangements should be made to have appropriate equipment available to deal with emergencies caused by fires, spills and poisoning.
- **Management.** Stocks of agricultural chemicals should be segregated by chemical type, i.e. herbicides separated from insecticides and fungicides etc. If some of the stock is not to be used shortly after purchase, ensure old stock is used first. Allow space in the store to keep empty packs whilst accumulating a disposal load. It is best practice to store fertilisers and stock foods in a separate store. Where only one store is available, it should be large enough so that fertilisers and stock foods are

kept well away from agricultural chemicals. Extra ventilation should be provided.

Protecting the environment

- Spraying should be carried out so that only the intended target receives the agricultural chemicals and any off-target movement is eliminated or minimised.
- Application of agricultural chemicals near watercourses such as rivers or streams requires special care to avoid any risk of water contamination unless this is an approved use.
- Care should also be taken to thoroughly clean all application equipment after use so agricultural chemical residues do not carry over when the equipment is next used.
- There should be a plan for disposal of any surplus spray and washings generated by the equipment cleaning.

Disposal of agricultural chemicals Agricultural chemical waste management

Three types of agricultural chemical waste cause problems:

- Liquid waste washings and leftover spray mixes.
- Packaging materials such as plastic drums and tins.
- Unwanted leftover agricultural chemicals.

Disposal of liquids

The following are some key points to consider when handling washings, leftover sprays and leftover product:

- If liquid is left over from a spray job and he same agricultural chemical is to be used immediately afterwards in the same spray tank, wherever possible, collect liquid and use it in the next spray mixture.
- After cleaning product containers, add the rinsings to the spray mix.
- Dispose of waste agricultural chemicals through the ChemClear program (http://www.chemclear.com.au/).
- Contact the agricultural chemical manufacturer for disposal details.

Storage of excess spray solutions are therefore best reused within the next few days only after the initial use or otherwise discarded.

Disposal of empty containers

- Reusable drums should be returned to the supplier.
- Cardboard packs can be composted or re cycled.
- Plastic drums should be triple rinsed and disposed of through drumMUSTER (an industry container recycling program) or local authority services.
- Containers with the drumMUSTER sticker can be taken to the nearest collection point after cleaning.

PPE should always be worn when handling or rinsing containers. The 1998 Queensland Farmers' Federation Environmental Code of Practices for Agriculture contains principles for good waste management (see Table 6.4).

Table 6.4 Waste management hierarchy

Classification	Practice	Objective
Least preferred	Waste disposal	The last waste management option that should only be adopted when all other practical possibilities have been considered. Waste should be confined to a defined area and managed to minimise harm to the environment or public health.
Good	Waste treatment	Using a process that turns waste into a form more easily used or more easily or safely disposed of (e.g. absorption by activated carbon, chemical or biological detoxification).
Better	Waste recycling	Strategies to reuse, reprocess, or recover a product (e.g. using returnable chemical containers; accessing programs that use crushed chemical drums for industrial fuel; spraying rinsewater on crop edges).
Even better	Waste reduction	Reducing the amount of waste created by using whatever means are available (i.e. changing practices; e.g. ultra low- volume applications; use of low-rate compounds).
Most preferred	Waste avoidance	Avoiding generation of waste at the source by looking at the situation and changing practices or by choosing the least hazardous products.

MANAGING AGRICULTURAL CHEMICAL RESISTANCE

Introduction

The repeated use of the same agricultural chemical tends to select (genetically) individual pests that are tolerant to that agricultural chemical compound. If the agricultural chemical is used frequently and repeatedly, populations can develop that are completely resistant to the applied agricultural chemical over several generations and sometimes the tolerance is carried over to related agricultural chemicals.

This section reviews the biological mode of action of agricultural chemicals and outlines how resistance can be managed. By placing agricultural chemicals into groups that work in the same way biologically, the repeated use of similar agricultural chemicals can be avoided. This reduces the chance of resistant individual pests being picked out in a pest population and delays the onset of resistance. The useful life of products currently registered can be prolonged by this strategy.

Resistance to agricultural chemicals

The onset of resistance in pests is a constant concern. Situations where pests are most likely to become resistant to agricultural chemicals are:

- where the pest population has a rapid turn-over and has been exposed to the same agricultural chemical for at least five successive generations
- in the case of plants-where ephemeral or annual species with prolific seeding rates, coupled with cross pollination, result in very large numbers of genetically varied individuals being exposed to similar agricultural chemicals

- where the agricultural chemical mode of action depends upon a single biochemical receptor.

Note

- The rapid spread of resistance is enhanced when plants have a high level of vegetative reproduction.
- Fungal resistance is relatively common especially following frequent use of systemic fungicides. This is probably due to the ability of fungi to rapidly reproduce and adapt to new conditions.

Examples of resistance in Australia

Herbicides

Resistant species	Herbicide affect
Wimmera ryegrass	diclofop-methyl
Wild oats	sethoxydim
Barley grass	paraquat
Cape weed	paraquat and diquat
Sow thistle, Indian hedge mustard, turnip weed, black bind weed	chlorsulfuron
Liverseed grass	atrazine

Insecticides

Resistance has been observed in many insect and mite families. It has been particularly common in the following instance:

- **Mites.** Many species of mite, especially two-spotted mite, have developed resistance to a wide range of miticides.
- **Flies.** Many flies, including blowflies and houseflies, have developed resistance to the synthetic pyrethroid group of insecticides.
- **Caterpillars.** Morethan 50 species of butterfly and moth caterpillars have developed some resistance to organopho-sohate insecticides. Similarly, resistance has developed agai-nst the synthetic pyrethroics applied to crops attacked by *Helicoverpa* spp.
- **Aphids.** Green peach aphids have developed resistance to organophosphates.

- **Beetles.** Lesser grain borers (beetles) have developed resistance to all organophosphates registered for use in stored gain.

Fungicides

In Australia, resistance has been reported with the use of a number of fungicidal compounds. As with weed and insect resistance, a range of resistance risk levels exist depending on the specific mode of action of the agricultural chemical. Systemic fungicides have the highest risk. Resistance to some systemic fungicides has been delayed by using them in mixtures with other non-systemic fungicides.

Managing resistance

To help manage pest resistance to herbicides, insecticides and fungicides, special letter codes have been introduced on agricultural chemical labels in Australia. These letter codes primarily indicate which mode of action each product possesses so that those agricultural chemicals that work in the same way can be grouped together. This enables agricultural chemical users to quickly distinguish those products that may complement one another in resistance management programmes.

In addition to letter codes, the organophosphate and the carbamate insecticide groups can be distinguished by the warning after the active constituent statement `an anti-cholinesterase compound'.

Herbicides

All herbicide labels carry a prominent capital letter of the alphabet indicating the mode of action grouping. There are 14 recognised biological modes of action among the registered herbicides, so the letters run from A to N. The risks of resistance arising when associated with each group vary from high to moderate to low. Details of the modes of action may be found in Appendix A.

Insecticides

Many of the insecticides that were first registered were either organophosphates or carbamates, both of which kill insects in a similar way. These two chemical families, together with the synthetic pyrethroids and the organochlorines, all attack and disrupt the insect nervous system. Details of insecticidal modes of action may be found in the tables in Appendices D and E.

Other insecticides affect the way insects change their outer covering as they grow larger, others interfere with the processes that change larvae into adults and others disrupt metabolic activities such as digestion.

Insecticides also include bio-control agents such as bacteria and viruses. Their groupings are listed in Appendix E.

Fungicides

As with herbicides and insecticides, it is a requirement that fungicide labels are marked with a letter of the alphabet indicating the mode of action group, which reflects the way they work. Two of the groups are entitled 'multi-site activity' and 'unspecified'. The scientific understanding of fungicidal biochemistry is continually developing and new groups are being added when new discoveries are made. An important development has been the discovery of the Strobilurin group, which is active on a wide range of fungal families. Details of modes of action may be found in Appendices F and G.

Summary

An agricultural chemical user should observe the following main points:

- Where a resistance management strategy has been established, follow it.
- Set up a recording system that provides the information needed for management decisions. This will include detailed monitoring data of pest populations so that agricultural chemicals are only used when economic thresholds are reached.
- Select agricultural chemicals on the basis of compatibility with beneficial species to avoid putting these 'beneficials' at risk.
- Do not use agricultural chemicals with similar modes of action in succession if this can be avoided.
- Do not spray areas larger than those needing treatment.
- Always follow the manufacturer's dose recommendation that appears on the label.
- Plan crop rotations where this is possible so that both winter and summer crops overlap. This is often a good way of eliminating pests such as weeds without overuse of selective agricultural chemicals.

- Use management practices aimed at reducing the seasonal carry over of pest populations, e.g. preventing weed seed-set, keeping fallows weed free and turning the soil to expose *Helicoverpa* spp. pupae to desiccation.

An integrated approach using as wide a variety of pest management techniques as possible is the basis for success. The aim is to minimise the number of times a pest population is exposed to an agricultural chemical, therefore reducing the opportunities for resistance to develop. Another aim is to reduce the exposure of pest populations to any agricultural chemical so that the opportunity for selection pressures is minimised. What is required is referred to as `Good Farming Practice' or 'Best Management Practice' and, as far as agricultural chemical selection is concerned, needs to be part of best practice.

With pest resistance in mind: 'When you are on a good thing, DON'T stick to it'.

SPRAY AND OTHER APPLICATION TECHNOLOGIES

Introduction

Users of agricultural chemicals face a problem: how can a small quantity of a biologically active chemical be applied evenly and efficiently over a crop or target area to control pest species?

Application methods can include:

- Seed treatment. The treatment of seeds with agricultural chemicals before planting, minimising seed damage or losses due to insects and diseases.
- Application of granules. Some insecticides and herbicides are applied as granules, for example, S-methoprene, a larvicide used for mosquito control, and hexazinone, a herbicide used for vegetation control.
- Stem or tree injection.
- Wick wiping. Wick wiping is particularly suited to systemic herbicides like glyphosate.
- Spray application.

This section relates specifically to spray application.

Spray application

The most common method of spraying agricultural chemicals involves creating a large number of small droplets from a liquid that contains the active ingredient at the correct concentration.

Typically, the agricultural chemical is formulated as a liquid, which allows the agricultural chemical user to transport the material to the crop and apply it to the target as a spray. The active ingredients of many agricultural chemicals are manufactured as crystals or viscous (sticky) liquids.

Successful spray application involves three phases:

1. **Droplet generation.** This creates a large number of droplets from a body of liquid.
2. **Droplet transmission.** The movement of droplets from the spray equipment (nozzle) through the air to the target.
3. **Droplet capture.** Droplets depositing on the target.

The aim of a successful spray application is to ensure that the correct quantity of agricultural chemical ends up on the intended target, with minimal or no contamination of off-target areas.

Droplet generation

Three methods for droplet generation are:

1. **Hydraulic.** This method forces the spray liquid, under pressure, through a small, specially designed orifice, or opening. Such devices are normally referred to as hydraulic nozzles. They are available in a number of different types and are designed for different purposes. The most common types of hydraulic nozzles used for spraying are hollow cone and flat fan nozzles. The air induction nozzle, a recent introduction to spraying, is an example of a modified hydraulic nozzle.
2. **Centrifugal force.** This method subjects the liquid to centrifugal force by precise feeding of liquid onto a spinning disc. Using this method, it is possible to generate droplets at the edge of the disc as liquid is spun off into space. Droplet size is influenced by the rotational speed of the disc, the nozzle design and liquid flow rate. Spinning cages, used in Micronair units, are an extension of this principle. The spinning cages are normally referred to as controlled droplet applicators (CDA), as they produce a narrower range of droplet sizes than hydraulic nozzles.
3. **Airshear.** This method feeds liquid into a high velocity jet of air. The mechanical impact of the moving air streams disperses the liquid into droplets. Systems employing this principle are normally referred to as airshear nozzles.

As the forces, described above, are applied to a spray solution, they change the physical shape of the liquid. Liquids do not compress easily, so when they are subjected to a force, they tend to change shape in react on to the value and direction of the force.

The phases of droplet formation are:

- **Sheet formation.** The appearance of liquid sheets through a hydraulic nozzle, or sometimes just beyond the edge of a spinning disc.
- **Ligament formation.** The formation of ligaments or 'strings' of liquid, often at the edge of the sheets.

In order to form droplets, the sheets and ligaments must disintegrate further. The sheets and ligaments usually disintegrate at the edge (rim disintegration) or centre (perforation) of a sheet, or through a wave motion that breaks up the liquid body. These modes of disintegration are shown in Figure 8.1.

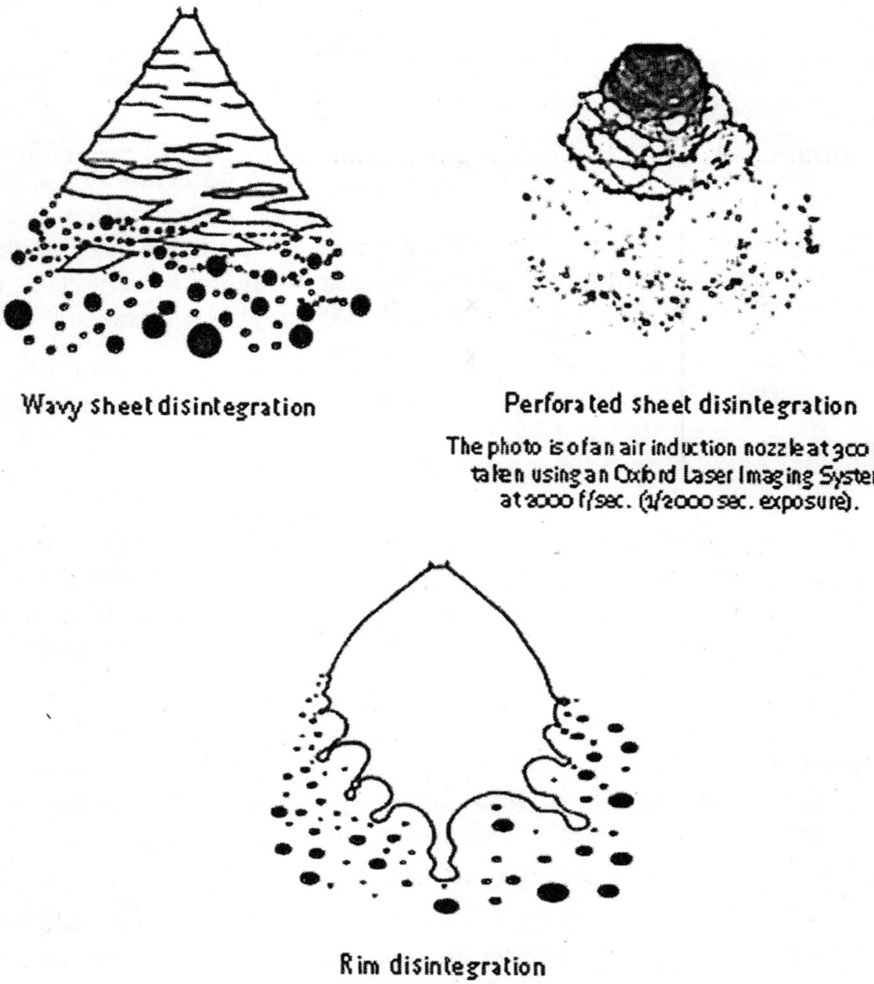

Figure 8.1. Diagrams of droplet formation from liquids (*Source: Pesticide application manual, 2nd edn.* Queensland Department of Primary Industries, Information Serves, Q189003, 1990)

Controlled droplet applicator (CDA) nozzles can produce droplets directly from a spinning disc (Figure 8.2). Usually, low flow rates are required to produce consistent droplet sizes. As flow rate increases, ligaments for n, which then break into droplets. At very high flow rates, flooding of the nozzle may occur, producing a sheet of liquid that disintegrates into droplets in the same way as hydraulic nozzles.

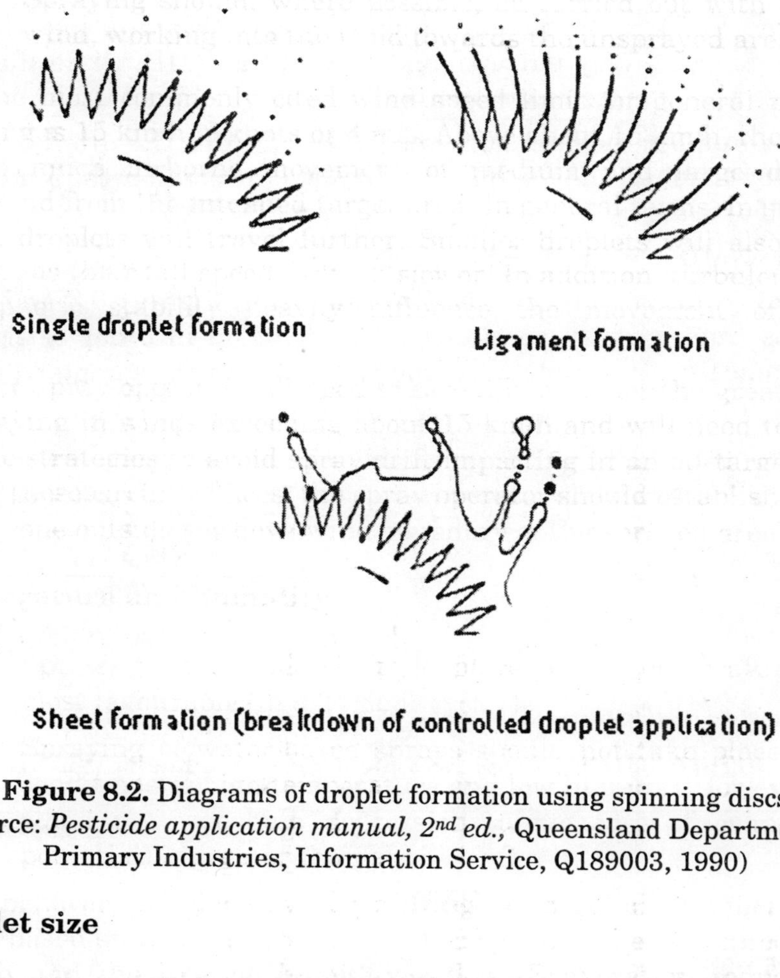

Figure 8.2. Diagrams of droplet formation using spinning discs (Source: *Pesticide application manual, 2nd ed.,* Queensland Department of Primary Industries, Information Service, Q189003, 1990)

Droplet size

Droplets used for spraying agricultural chemicals are small. Droplets are measured in microns (μm).

As an example, the full stop at the end of this sentence is approximately 300 pm in diameter. A micrometre (μm) is 1/1000 of a millimetre (mm). Therefore, a 500 μm droplet is half a millimetre in diameter. A 500 μm diameter droplet is considered a large droplet in spray application technology.

All nozzles produce a range of droplet sizes. It is therefore difficult to accurately describe the output from a spray nozzle. Many agricultural chemical labels describe the droplet size in terms of the volume median diameter (VMD).

The VMD divides the droplet range into two equal parts. Half of the total volume is in droplets larger than the VMD and half is in droplets smaller than the VMD. Two different nozzles may produce the same VMD but a different droplet range. One produces droplets that all fall in a very narrow band around the VMD, while the other produces a very large range of droplet sizes.

The number median diameter (NMD) divides the droplet sample in half by number, that is, half the droplets are smaller than the NMD and half are larger.

The NMD is always less than the VMD.

A diagram of VMD and NMD is shown in Figures 8.3 and 8.4. If droplets from a spray nozzle could be lined up in order of size, the VMD indicates the droplet size that would divide the sample in half by volume.

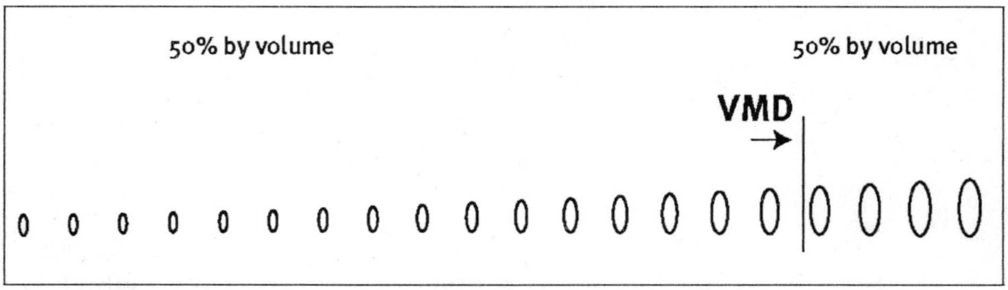

Figure 8.3. Diagram of the volume median diameter (VMD)

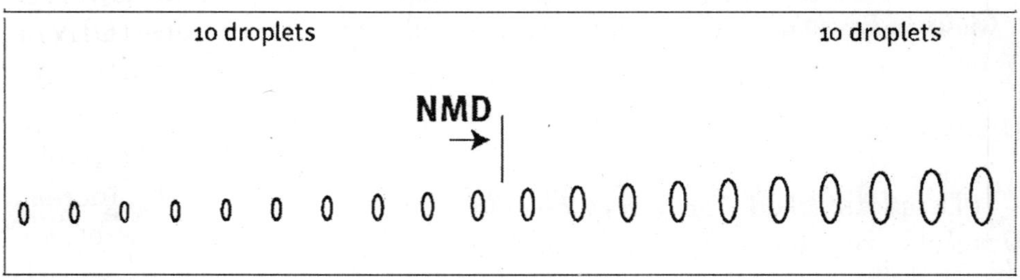

Figure 8.4. Diagram of the number median diameter (NMD)

Droplet size is a very important spray characteristic. The larger the droplet diameter, the less a spray droplet is at risk of drift (spray drift is discussed in Section 10). Droplet size, initial speed of travel, rate of evaporation and meteorological conditions all influence the movement of a droplet through the air.

Droplet size, volume and coverage

There is an important relationship between the size, volume and number of droplets that are produced in a fixed volume of agricultural chemical.

Simple relationships show that as the size of droplets produced by a nozzle is halved, the number of droplets produced from a fixed volume of agricultural chemical increases by a factor of eight. This calculated relationship is illustrated in Table 8.1.

Table 8.1 The relationship between droplet number, volume and ground coverage

Droplet size (diameter)	Number of droplets per mL	Number of droplets deposited on 1cm² at 1L/ha
20	239 000 000	2390
80	3 730 000	37.3
100	1 910 000	19.1
200	239 000	2.39
400	298 00	0.30
1000	1910	0.02

For the same spray application volume, as droplet size is decreased, larger numbers of droplets are produced, which can improve the coverage of targets. Often carrier volumes (the amount of oil or water used to mix the agricultural chemical) can be lowered as the droplet size is reduced. Ultra low volume formulations (ULV) are usually applied in volumes as low as 2 L/ha by using nozzles with VMDs of about 50-80 mm.

On the other hand, as droplet size is increased (which is required to reduce spray drift), carrier or water volumes are usually increased to maintain coverage of a target, meaning greater amounts of water must be carried in the spray tank.

Droplet transmission

Once the droplet is produced, it must be transferred through the air to the target. A droplet's movement through the air from a nozzle to the target is influenced by the droplet's size, initial speed of travel and rate of evaporation, as well as meteorological conditions.

Most ground application equipment releases droplets close to a target canopy. Therefore, the initial speed from a spray nozzle and the air assistance can influence the transmission of droplets over short distances (less than 100 cm).

However, with air-assisted spraying, motion soon slows and droplets assume the speed of the air around them. The droplets quickly move entirely under the influence of gravity and the local prevailing wind and turbulence.

Gravity causes droplets to fall towards the earth's surface. This is referred to as sedimentation. When a balance occurs between air resistance and the pull of gravity, the speed reached by a droplet's fall is known as its sedimentation velocity (or terminal velocity). This varies with the diameter of the droplet, being low for small droplets and greater for large droplets.

The speed of a droplet falling towards the ground is very important. The longer a droplet takes to fall, the more time is available for it to be shifted away from its intended target by cross winds. To complicate matters further, low humidity and high temperatures can cause small droplets, particularly water-based droplets, to shrink by evaporation and remain at risk of sideways movement for longer time periods.

Large droplets, because of their greater mass and sedimentation velocity, tend to land quickly on the target or the ground close to their point of release. Small droplets, however, fall slowly, so a prevailing wind could move them away from the treatment area.

Table 8.2 illustrates the fall speeds and possible downwind distances that droplets will be transported to if released three metres above a crop in a steady cross wind, blowing at one metre per second. This table does not include the effects of evaporation and turbulence.

In practice, effects such as turbulence have a major influence on downwind deposition. Table 8.2 should not be used to predict safe downwind buffer areas.

Notice in Table 8.2 how the small droplets fall very slowly through the air and are blown long distances, while the larger droplets are pulled towards the ground more quickly and consequently only travel short distances away from their intended target.

Table 8.2 Sedimentation velocity for different droplet sizes

Droplet diameter (μm)	Sedimentation velocity (m/s)	Time to fall 3 m	Metres displaced downwind in 1m/s wind
1	0.00003	28.1 hours	10000
10	0.003	16.9 minutes	1000
20	0.012	4.2 minutes	250
50	0.075	40.5 seconds	40
100	0.28	10.9 seconds	10.7
200	0.72	4.2 seconds	4.2
500	2.14	1.7 seconds	1.4
1000	5.0	0.8 seconds	0.6

As droplet size decreases, the movement of air around a crop or target area controls how the spray is spread out. Small droplets tend to diffuse in the atmosphere, scattering downwind under the influence of both horizontal and vertical movement.

Consequently, it is important to understand the way in which air is moving above a crop in order to spray effectively and control the off-target movement of agricultural chemicals.

Droplet capture

For spraying to be effective, droplets must land on the plant or insect target. The size of a target, its shape and orientation as well as droplet velocity, size and weather conditions all influence the ability of the droplet to land on the target.

Objects, such as branches and leaves on crops, cause air to deflect (or move) around them. Small droplets tend to follow the path of air surrounding them, whereas larger droplets can have enough drive to move away from this air and hit the ta·get. This is shown in Figure 8.5.

The ability of a target to catch spray droplets is called the etching efficiency. The greater the catching efficiency, the more chance there is of spray landing on a target.

In general, the catching efficiency increases as the:

- speed of a droplet moving towards the target increases
- diameter or width of the target decreases
- droplet size increases.

Therefore, the catching efficiency of a plant or insect increases as it becomes smaller. For this reason, small hairs on insects and items or needle-like leaves on trees (e.g. *Casuarina* spp.) tend to be more efficient at catching small droplets than broad smooth leaves (e.g. cotton leaves). This is illustrated in Figure 8.6.

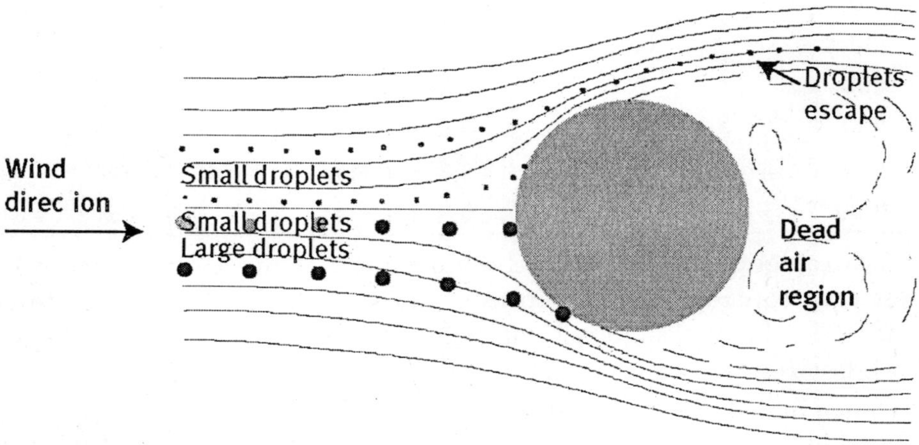

Figure 8.5. A diagram showing the effect of target dimension on droplet capture

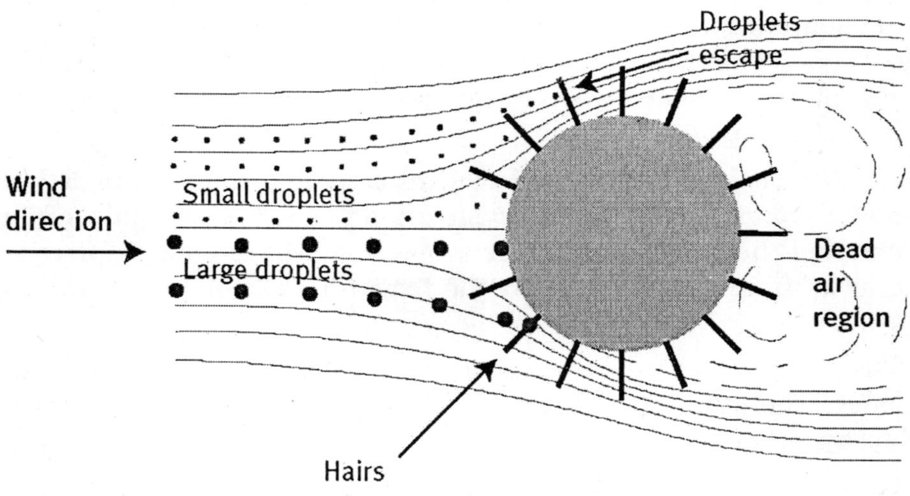

Figure 8.6. A diagram showing the effect of target dimension (fine hairs) on droplet capture.

Note : In Figure 8.5, the target does not catch some small droplets. In Figure 8.6, all are caught.

Spray nozzles

Nozzle classification schemes

To help applicators select nozzles appropriate for particular agricultural chemicals and circumstances, international standards have been developed to define spray quality in a more practical way.

XR TeeJet® (XR)

	bar						
	1	1.5	2	2.5	3	3.5	4
XR8001	M	F	F	F	F	F	F
XR80015	M	M	M	F	F	F	F
XR8002	M	M	M	M	M	F	F
XR8003	M	M	M	M	M	M	M
XR8004	C	C	C	M	M	M	M
XR8005	C	C	C	C	C	M	M
XR8006	C	C	C	C	C	C	C
XR8008	VC	C	C	C	C	C	C
XR11001	F	F	F	VF	VF	VF	VF
XR110015	F	F	F	F	F	VF	VF
XR11002	M	F	F	F	F	F	F
XR11003	M	M	M	F	F	F	F
XR11004	M	M	M	M	F	F	F
XR11005	M	M	M	M	M	M	F
XR11006	C	M	M	M	M	M	M
XR11008	C	C	M	M	M	M	M

VF – Very Fine
F – Fine
M – Medium
C – Coarse
VC – Very Coarse
XC – Extremely Coarse

Figure 8.7. An example of a classification scheme used by `Spraying Systems' to define spray quality (based on ASAE S_{572}). Note: Bar is an expression of pressure. One bar is atmospheric pressure at sea level.

A nozzle classification scheme was created in the mid 1980s to define and describe the measurement systems and droplet range that a spray nozzle produces.

Five categories of spray quality were devisee: very fine, fine, medium, coarse and very coarse. The American system (ASAE S572) includes the additional category of extra, coarse. The scheme was originally developed for defining, ground hydraulic application nozzles; however, the scheme now includes other nozzle types. Industry defines the categories using four reference sprays (defined nozzles and pressures).

Such classification systems enable regulators, researchers, agricultural chemical users and growers to standardise the description of nozzle systems and therefore spray quality. Figure 8.7 shows an example of a classification scheme used by a commercial supplier of nozzles. Nozzles produce fine to very coarse droplets depending on the orifice size and spray pressure.

Hydraulic nozzles

Hydraulic nozzles produce droplets by forcing the agricultural chemical liquid, under pressure, through a small, specially designed orifice, or opening.

There is a wide range of nozzles available for many different tasks. It is important that the most appropriate nozzle is selected for each task. All users should carefully read the manufacturer's specifications and technical information before selecting and using hydraulic nozzles, including instructions on their care and maintenance.

A hydraulic nozzle performs four important tasks:

1. It produces droplets of a known specification.
2. It provides direction to the liquid.
3. It creates a pre-determined spray pattern.
4. It acts as a metering device by allowing (at a set pressure) the flow of a set amount of liquid.

In general, the size of droplets from hydraulic nozzles in still air becomes smaller as:

- pressure increases
- orifice size decreases
- fan angle (flat fan nozzles) increases.

Hydraulic nozzles should be carefully looked after, as instructed by the manufacturer, and replaced when worn. Worn nozzles can produce uneven patterns and incorrect droplet sizes. In addition, worn nozzles produce higher flow rates that can lead to significant errors in calibration. Finally, they can cause errors in the amount of product delivered to the target.

Nozzles are manufactured from a variety of materials. Plastic, stainless steel and ceramic nozzle tips are the most common. Nylon and plastic nozzles are inexpensive and suitable for most situations. Where harsh agricultural chemicals are used, under high pressures, ceramic nozzles are often favoured, as they are more resistant to wear.

Hydraulic nozzles should be checked regular y for damage, blockage and flow rate. Nozzles are to be replaced when the output varies outside the manufacturer's recommended flow rate by more than 10%.

Types of Hydraulic nozzles

The main types of hydraulic nozzles are:

Flat fan (tapered). Tapered, flat fan nozzles are designed to deliver more spray to the centre of the nozzle. These nozzles are primarily designed for boom spraying units, enabling sprays from adjacent nozzles to be overlapped. They are usually operated between 200 and 400 kPa.

Flat fan (low pressure). Low-pressure, flat fan nozzles are similar to tapered nozzles, but they are designed to operate at lower pressures. The lower pressures reduce the production of small droplets, which are prone to drift.

Flat fan (even spray). Unlike tapered, flat fan nozzles, these nozzles give a consistent deposit across the width of the spray plume. They are primarily designed for single nozzle operation, for example, when using knapsack spraying units.

Hollow cone. These nozzles produce a hollow ring of spray and are usually operated at higher pressures. For example, a pressure greater than about 500 kPa will produce fine droplets. These nozzles are designed mainly for foliage spraying and far use on orchard spraying units (see Section 9). Most hollow cone nozzles use a swirl plate to rotate the spray in a swirl chamber before exiting through a narrow orifice. The swirl plate may be a separate part or incorporated into the tip of the nozzle.

Solid cone. By forming a hole in the centre of the swirl plate, the solid cone nozzle can make the spray fill the centre of a hollow cone spray plume. Generally, this assists the application of higher flow rates and a coarser droplet range.

Flood jet (anvil). These nozzles produce large droplets at high flow rates. They produce a wide angle at low pressures. The flood jet nozzle and knapsack spraying units can apply larger (herbicide) droplets.

Air induction and injection nozzles. These nozzles use a mechanism to introduce air into the spray mixture prior to the release through the orifice and droplet formation. Air is usually drawn into the nozzle body using small (venturi) holes. A mixture of air and spray droplets is released by the nozzle, which produces a coarse droplet range.

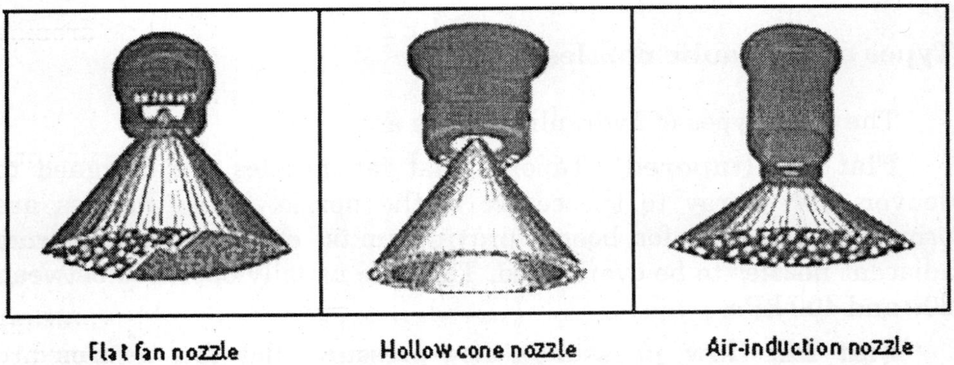

Figure 8.8. Examples of hydraulic nozzles

The advantages and disadvantages of hydraulic nozzles

Advantages :

- Nozzles are easy to use-no moving parts.
- Nozzles can be used for a wide range of situations.
- Nozzle components are easily changed.
- Nozzle parts are simply designed.

Disadvantages:

- Nozzles are prone to wear.
- Nozzles require regular calibration.

- Knowledge of operational parameters is required (e.g. pressure and distance to target).
- Air assistance is sometimes required for efficient small droplet capture.
- Nozzles generate a wide droplet range that can lead to wastage (large droplets account for a large carrier volume) and very small droplets contribute to spray drift.

Controlled droplet application (CDA) nozzles

General principles

If liquid is supplied to the centre of a rotating surface, centrifugal force spreads the liquid to the edge of the surface and droplets are formed (see Figure 8.2).

As liquid flow rate is increased, three distinct phases of droplet formation are observed:

1. single droplet formation
2. the formation of long strands or ligaments that break up into droplets
3. the formation of a sheet of liquid that breaks up and disintegrates into ligaments and droplets.

In the third case, sheet formation occurs when the rotating surface is flooded with liquid. A wide range of droplet sizes is formed.

However, the first two processes are capable of producing precise spectra that are primarily dependent on the rotational speed of the surface. So, when the surface rotation becomes faster, the droplets become smaller.

If the outer edge of rotating discs are equipped with teeth and grooves are used to feed the liquid to the edge of the disc, narrow ranges of droplet sizes can be formed, as long as liquid flow rates are not too excessive.

This principle underlies the design of spraying units, such as the hand-held, battery-operated 'Micron Herbi'. Designed to produce droplets between about 200 and 250 µm, this nozzle system is best used for applying coarse droplets in herbicide application. Such nozzles tend to produce lower amounts of fine droplets.

CDA nozzle systems have also been developed for insecticide and fungicide application. However, despite their superior ability to control and regulate droplet size, they have not been widely adopted in commercial spray systems.

The advantages and disadvantages of CDA technology

Advantages:

- A narrow range of droplets can be produced (compared to hydraulic nozzles).
- Droplet size can be controlled and the formation of unwanted (large) droplets minimised, so carrier volumes can sometimes be reduced.
- Nozzle systems can be tailored for producing certain droplet sizes. For example, high-speed, small discs produce fine droplets; low-speed, large discs produce larger droplets.

Disadvantages:

- Most nozzle systems require relatively complex motorised components. Therefore, maintenance can be high and durability low.
- The effective use of such equipment requires specialist knowledge and a high level of understanding by the user.
- Accurate and correct droplet formation requires a precise balance of disc design, rotational speed and liquid flow rate.

Airshear nozzle systems General principles

Airshear nozzle systems produce droplets by subjecting a liquid stream of agricultural chemical into a high velocity air stream. The action of the air upon the liquid causes the liquid to be broken up into a large number of small droplets. Liquid pressure, or orifice design, is not used to produce the droplets.

Petrol driven engines, motors or power take off (PTO) of tractors are used to drive fans, which direct air over the nozzle orifice. Normally, air speeds in excess of 300 km/h are required to produce small droplets. The use of low air speeds prevents correct atomisation (liquid break up) occurring at the nozzle orifice. It can result in large droplets forming.

Deflectors, diffusers and electrostatic charging equipment are sometimes used in conjunction with airshear systems. Airshear nozzles are found on back misters as well as tractor-drawn, air blast misting equipment.

The advantages and disadvantages of airshear nozzle systems

Advantages:

- Lower volumes of carrier (e.g. water) can normally be used.
- If the air speed is high enough, a relatively narrow range of small droplets can be produced.
- Airshear systems can move droplets into plant canopies. They use air movement to enable the droplets to penetrate and settle on the canopy.

Disadvantages:

- Not suitable for herbicide application.
- Careful calibration is required to match the crop with the spraying unit output.
- Air speed and liquid flow rates are critical for proper droplet formation.
- Small droplets can move away from the target area.

Weather and spraying

Spray droplets, particularly small droplets, move with the air that surrounds them. The prevailing weather conditions at the time of spraying greatly influence spray droplet spread and capture. Weather conditions that influence the application of agricultural chemicals are discussed below.

Horizontal wind speed and direction

The sun heating the ground and sea causes air masses of varying temperature, humidity and air pressure to develop over the surface of the earth. Air pressure gradients are established, which cause air to flow from high to low-pressure areas around the globe. Except near the equator, forces, due to the curved shape and rotation of the earth, cause the air to flow anticlockwise (in the southern hemisphere) ground high-pressure areas (anticyclones) and clockwise around low-pressure areas. This flow of air is wind.

It is important for a user to know the wind speed and direction above a crop or target area before carrying out any spraying operation. Not only does the wind largely control the direction of spray after release from a nozzle (and therefo˜e its movement towards or away from the target area), but also the degree to which droplets are caught by the foliage or pest, as well as the evenness of the spray deposit.

Turbulence

Turbulence can develop over the ground as a result of the thermal (upward) movement of warm air or the mechanical movement of wind across the ground. A wind or breeze travelling close to the surface of the earth rarely has a smooth flow. Instead, the atmosphere produces a turbulent motion in the air caused, in part, by the movement of air layers against each other, as well as frictional losses of energy at the earth's surface. This effect is shown in Figure 8.9.

The lower air layers move less than the upper layers because their energy is Lost at the earth's surface. Figure 8.9 also shows that close to the earth's surface, the average wind speed usually increases with height.

The extent of this turbulence is also determined by the roughness of the surface. For example, a stand of trees or a tall crop tends to produce greater turbulence for a given wind speed than an area of mown grass.

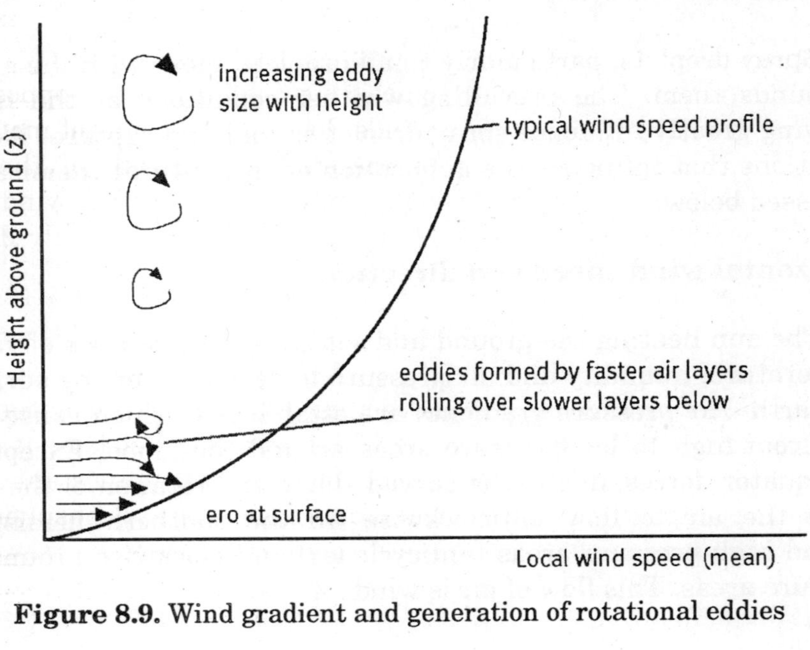

Figure 8.9. Wind gradient and generation of rotational eddies

Local wind effects

Although primary wind directions end wind speeds are caused by large-scale systems, small-scale local winds can be produced, which can have a significant impact on spraying. Some examples are given below:

- **Thermal movement of air.** In large open areas, under conditions of high radiation, unequal heating of the ground can occur. The inflow of cooler, heavier air displaces the warmer (less dense) air, causing it to rise. Under such conditions, large-scale circulation currents can be formed with resulting wind flows and turbulence. On a small scale, the difference in temperature between a crop and fallow (bare soil) areas can also give rise to local air currents around and within a crop canopy.
- **Thunderstorms.** Thunderstorms can produce strong winds in all directions. Although storms generally track west to east across eastern Australia with the passage of frontal systems (in the south), local wind directions can be highly variable.
- **Katabatic winds.** In hill country or on flat land close to slopes, farming areas can be subject to evening katabatic winds. As the land cools, air immediately above the surface can be cooled resulting in that air becoming denser and therefore heavier than the surrounding air. If cooling occurs on sloping ground, heavy air can flow, under gravity, to lower levels, resulting in the production of local wind flows.
- **Anabatic winds.** Anabatic winds are the opposite of katabatic winds. These winds flow up the slopes of mountains from the valleys due to radiational warming of the lower slopes and valleys causing air to rise. An anabatic wind is also named a valley breeze. Anabatic winds are typically daytime winds during the summer.

Vertical air movement (stability)

As well as horizontal air movement, the vertical displacement of air has to be taken into account while spraying. In fact, it can be argued that this parameter is one of the most important factors influencing agricultural chemical application. The atmosphere is a three-dimensional space in which air and droplets can be transported vertically as well as horizontally. The temperature of air it the lower atmosphere decreases with height.

A parcel of air displaced upwards from the ground (e.g. by convective thermal effects) will normally move into an area of lower pressure and expand. As it expands, the air parcel cools. The rate of cooling is about 10^{0}C per 1000 metres.

In normal summer conditions during the late morning and afternoon, air parcels that are produced tend to rise and remain hotter, and therefore lighter, than the surrounding air. The air is said to be unstable and is characterised, if there is enough moisture in the atmosphere, by the formation of large, cumulus clouds. Air made to rise under such conditions tends to continue its upward motion.

At other times, the atmosphere is stable. Under such conditions, a parcel of air is cooler, and therefore denser, than its surroundings and tends to return to its original position before displacement. Vertical air movement tends to be suppressed and wind velocities are usually low.

A stable atmosphere can occur on dry, cloudless nights when the land cools as the ground emits long wave radiation (heat energy). The ground then cools the air above it and a surface temperature inversion occurs. Under these conditions, the temperature increases with height above the ground over a short distance, before resuming normal conditions.

Air movement and spraying

The above explanation of vertical and horizontal air movement can be used to understand the droplet transport process. When droplets are large (greater than 300 μm), their passage towards the ground from a spraying unit is largely influenced by gravity. In other words, the droplets tend to hit the ground largely unaffected by air currents, unless the wind velocities are very large. However, most spray equipment produces a proportion of small droplets (smaller than 150 µm). As droplets decrease in size, the movement of air around them increasingly controls droplet movement (see Section 9 on drift management).

Consequently, it is important to understand the way air moves above a crop in order to spray effectively and control the off-target movement of material.

Temperature

When water-based droplets are released into the air by a spraying unit, they become smaller as water evaporates from them. Therefore droplet size can reduce significantly after the droplets have left the nozzle.

The problem is particularly serious for small droplets for the following reasons:

- As the size of a droplet decreases, the ratio between the surface area of a droplet and its volume increases. This means that a greater proportion of the volume of the droplet is exposed to the atmosphere as the droplet size decreases.
- As a droplet becomes smaller through evaporation, its sedimentation velocity (fall speed towards the ground) becomes slower. So, as a droplet remains airborne longer, there is further evaporation and it becomes even smaller.

Researchers have calculated the lifetime of droplets and the possible distance they will fall under the influence of gravity. This work has shown that evaporation increases as the temperature increases and the air becomes drier (relative humidity decreases).

Water is the main material used to dilute agricultural chemicals. As it evaporates, extreme care must be taken when applying small. droplets. The drift potential of a water-based spray consisting of large droplets may increase under conditions of high temperature and low humidity. Ultra low volume (ULV) agricultural chemicals (see Section 4) attempt to overcome the impact of evaporation by using mineral and vegetable oil carriers.

Relative humidity

Relative humidity (RH) is used to describe the dryness of the atmosphere. It defines the ratio of the amount of water that is contained in a sample of air to that which could be contained in the same volume of air if saturated at that temperature. Because it is a relative measure dependent upon temperature, the RH increases as the temperature drops, and decreases as temperature rises for air with the same moisture content. Therefore, it is usual to find that over a crop canopy, maximum RH values are recorded at dawn, because dawn is generally the coldest time of the day.

Monitoring the weather

Given the importance of weather conditions on spray application, it is important to get an accurate weather forecast before application. One of the best and most up-to-date methods for obtaining a weather forecast is to access the Bureau of Meteorology website. This site provides current regional forecasts satellite images and weather charts. Such information can greatly assist the planning of a spray operation. Go to http://www.bom.gov.au

However, it is important to also observe and record the weather conditions in the spray area both before and during spraying and note any changes. Not only can a forecast be inaccurate, but spray application is primarily affected by the local conditions existing at the spray site during application.

As a minimum, wind direction, wind speed, temperature and relative humidity should be recorded. Small, hand-held devices are available for measuring wind speed, temperature and humidity. Hand-held battery operated anemometers (wind meters) are inexpensive. Some units also contain solid-state transducers to measure temperature and relative humidity at the same time. The information is usually displayed on a large, clear LCD display.

Where spray application occurs regularly over a larger area, consider purchasing portable weather stations which are capable of logging and storing the weather information for later retrieval and analysis by computer.

SPRAY APPLICATION EQUIPMENT

Introduction

The previous section explained how droplets are formed and transported in the environment. This section explains how spray equipment is designed, calibrated and operated. Spraying units come in a large range of types and sizes, from small, hand-held spraying units to large self-propelled machines.

Types of spraying units

Boom spraying units

These units are commonly used for applying agricultural chemicals to held and vegetable crops. Hydraulic nozzles are the most common form of nozzle used on boom spraying units, although controlled droplet applicators (CDA), airshear, twin fluid and electrostatic nozzles are also used.

Boom spraying units come in many different designs. In general, they can be grouped into four configurations: tractor-mounted, vehicle-mounted, fully-trailed and self-propelled spraying units.

Tractor-mounted boom spraying units

Tractor-mounted boom spraying units are commonly used for spraying small fields, or difficult or hilly land. This style of spraying unit is very compact, highly manoeuvrable and requires a minimal capital outlay, as most farms already have a suitable tractor. Tank sizes vary from about 500 to 1200 L, depending on farm size and the size of the available tractor. Generally, spray boom lengths of 10 to 24 m are used.

Vehicle-mounted boom spraying units

Vehicle-mounted boom spraying units have become more popular in recent years due to their increased speed, comfort and versatility. These units can be quickly driven to paddocks, easily transported between treatment areas even when they are considerable distances apart, and quickly returned to filling and loading stations. Often, 4WD motorbikes, utilities and trucks are used to carry this spraying unit (see Figure 9.1).

Figure 9.1. Vehicle-mounted boom spraying unit

Fully-trailed boom spraying units

Fully-trailed boom spraying units have all the spraying unit components, including the pump. The tank and boom are mounted on an independent chassis or frame. The units can be towed by anything from a four-wheel motorbike to a truck. Trailed units can usually carry a heavier load (for a given vehicle size), as she load is distributed over three or more axles. They are also quicker to hitch and release than mounted units, so the towing vehicle can be readily used for other tasks.

Self-propelled boom spraying units

Self-propelled boom spraying units are purpose-built for spray application and do not have to compete with general-purpose activities that require tractors and utilities. Spray equipment does not have to be routinely fitted or removed from this type of unit, so they can be quickly moved into the field for spray operations. The spraying units can be designed and set up for a variety of spraying needs. For example, high clearance axles can be used for tall crops, while fitting the unit with low-pressure tires is suitable for operations over soft ground. Designed with air-conditioned, closed cabs, the spraying units are also generally comfortable to operate and can be driven at higher speeds than tractors. These units require a large capital outlay and have therefore, traditionally, only been operated by spray contractors and large agricultural operations (see Figure 9.2).

Figure 9.2. Self-propelled boom spraying unit

Hand-held spraying units

In Australia, hand-held spraying units are normally only used for small-scale and infrequent use operations. However, some smaller, intensive operations, such as nurseries, regularly use hand-held spraying units. The most common type of hand application equipment is the knapsack spraying unit. Knapsack spraying units consist of a tank designed to be worn on the user's back, as well as a hand-operated piston or diaphragm pump and lance. On some types, a small motor is used to operate the pump.

Figure 9.3. Hand-held compression spraying unit.

Another common type of hand spraying unit is the compression spraying unit. These spraying units use an air pump to create pressure in the spray tank. This produces liquid flow and a pressure head (see Figure 9.3).

A major disadvantage with hand-pump spraying units is that the pressure produced by the hydraulic nozzles creates different flow rates and droplet sizes, according to how strongly the pump is operated. This problem can be overcame by fitting a valve such as the SMV (spray management valve) or CF Valve that is designed to control and maintain a steady flow rate, even if pressure changes.

Adjustable, gun-type nozzles are another common type of hand-held spraying unit. The nozzle is usually connected by a long length of hose to the remainder of the spraying unit, which may stand on a trolley or vehicle.

Hand-held spraying units with spinning discs (CDA spraying units) are also available.

Orchard spraying units

Orchard spraying units come in a wide range. Custom-built and commercial orchard spraying units are used for crop protection strategies in vine crops, evergreen and deciduous trees. The range of crop management techniques impacts upon the type of spraying unit that can be used and also the effectiveness of the spraying unit.

Most spraying units use the air to help carry droplets to the target zone within the tree canopy. Some spraying unit types use high velocity air to produce and transport droplets to the target.

Effectiveness of spraying units

The biological effectiveness and off-target loses that a spraying unit produces are determined by a number of different factors:

Spraying unit configurations

There are many types of spraying units used for applying agricultural chemicals in orchards. These include low profile, tower, air blast and combination units as well as units that use multi-ad spray systems and ducted conveyors. The spraying units may also be single or double-sided.

Nozzle types and pressure

A wide range of nozzles is used on orchard spray equipment. Ceramic hollow cone and solid cone (operational pressure range 5 to 20 bar) are the most common form of hydraulic nozzle. Airshear technology (200 to 350 km/h at the nozzle outlet) and CDA systems (X1, spanspray, Micronair) are also used.

Air volumes

A wide range of fan types and sizes, as well as a varying number of fans, are used on orchard spraying units. The location of the fan in relation to the tree canopy and the spraying unit can also influence air flow. Air calibration and configuration has a large influence on droplet movement and canopy penetration. Australian air blast equipment is normally calibrated to deliver between 15000 and 15000 m^3/h. Air volume should normally be matched to tree volume. If air volume is too low, droplets may fail to penetrate the tree properly, and so, coverage and effectiveness is reduced. If air volume is too great, droplets tend to pass through the foliage, increasing the drift potential.

Water volumes applied

Carrier water volumes vary greatly end depend upon equipment, industry, location and pests targeted. Volumes may range from 150 to 10000 L/ha.

Row and tree spacing

Row and tree spacing depend on the crop. There is a trend in some industries for high density planting of trees, that is, closer rows and trees within a row. Row spacings vary from 2.5 m to 12 m and tree spacings vary from 1.5 to 10 m or more.

Travel speeds

Travel speeds vary considerably and can range from 1.5 to 10 km/h. Ground speed does not usually have a large impact on spray drift.

Tree configurations

Allowing trees to form hedges is becoming common practice in many industries. This helps in spray drift management, as trees or the outer edge of a plantation can be used as vegetative barriers. However, reduced spray coverage on plantation boundaries can cause production problems.

Other application equipment

Wiper applicators are used to apply Herbicides directly to weeds with a herbicide-soaked sponge or wick. This can be a very effective method where there is a separation in height between the weed and crop or pasture. As no droplets are produced, drift from these applications is very low.

Spraying unit components

While there is a large variety of components that make up a spraying unit, there are some basic components that are found on most spraying units.

The basic components of liquid application systems include the following (not every spraying unit will have all the components listed):

- tank to hold the spray solution
- method to make liquid flow, most commonly a pump
- agitation system to mix agricultural chemicals and prevent settling during operation
- pressure regulators and control valves
- filtration system to prevent nozzle blockage
- method of producing droplets, most commonly a nozzle
- frame or chassis and drive
- auxiliary equipment such as clean water and agricultural chemical handling equipment.

A typical layout for ground application equipment is illustrated in Figure 9.4.

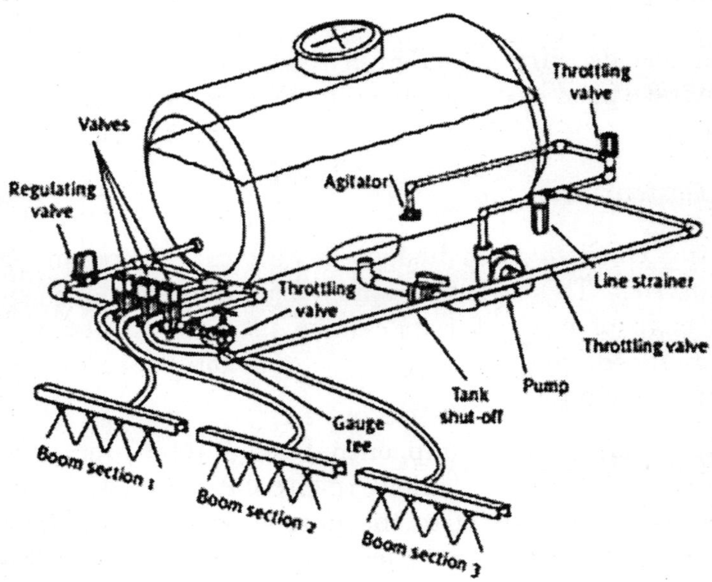

Figure 9.4. Common boom spraying unit components

Tank

The spray tank should be an appropriate size for the type of spraying unit that will be used. The spray should also be able to hold the volume required for the area that will be sprayed. The shape of the tank should allow easy access for filling and draining. A small sump in the tank is generally recommended, so that a minimum amount of liquid remains within the tank when the pump starts to pump air.

Materials used in construction need to be resistant to chemicals: non-corrosive, not easily damaged, resistant to sunlight and easy to repair.

Liquid flow

The liquid is usually forced from the spray tank to the nozzle by a pump, but some handheld systems use gravity for the liquid to flow (i.e. tank positioned higher than the nozzle). A wide variety of pumps are available for application systems. Examples include diaphragm, centrifugal, piston, roller and gear pumps.

When selecting a pump, the following factors should be considered:

- operating pressure
- output (L/min) of liquid
- power and methods available to drive the pump
- type of agricultural chemical to be used
- durability of the pump
- costs

Agitation system

Many agricultural chemical formulations consist of fine powders or particles that need to be held in suspension in the spray tank. If the mix is left to stand, these particles tend to settle on the bottom of the tank. A system to agitate or mix the agricultural chemical is therefore required. This is usually achieved by re-circulating the spray mix back to the spray tank. The pump output should be greater than that required by the nozzle to allow re-circulation back into the tank during spraying. Once flow to the nozzles is stopped, the total flow from the pump is redirected back into the spray tank. Sometimes, special nozzles are used to increase the agitation in the tank.

Other methods of agitation include mechanical systems such as a rotating paddle or manual agitation (physically shaking the task) for small, hand-held equipment.

Pressure regulators and control valves

Liquid flow rate and pressure to the nozzles must be controlled to ensure that the spraying unit output is consistent. This is generally achieved by use of pressure regulators and control valves. These may be operated manually, or by electronic controllers (particularly with the larger systems).

All systems must be fitted with a pressure gauge. The gauge should be positioned as close as possible to the nozzles.

Filtration system

Filters are required to prevent nozzle blockage. Blockage results in wasted time, increased risk of agricultural chemical exposure (if nozzles or filters require cleaning in the field) and poor coverage in the field if blockages are not detected.

Factors such as the water source, formulation and pump agitation capability influence the type of filtration system required for the spraying unit.

There are typically several stages of filtration in liquid application equipment. These stages and typical mesh sizes are listed below. Mesh size is defined as the number of openings along a linear inch. Therefore 100 mesh has 100 openings along a linear inch, or 10000 openings per square inch.

Filter stage	Typical mesh size
Tank inlet filter	50 mesh
Suction filter	40 to 80 mesh
Pressure line filter	40 to 80 mesh
Nozzle filter	50 to 100 mesh

Calibration

Importance of calibration

The objective, when applying agricultural chemicals, is to deliver the required amount of active ingredient in the agricultural chemical to the desired target area. Regular calibration allows the user to check that all components of the spraying unit are operating within acceptable limits. It will also reduce the risk of over-application, or applying rates that reduce efficacy to the target areas. Finally, it will reduce unnecessary contamination of the environment.

Over-application of agricultural chemicals occurs when more than the recommended amount of chemical is applied to the target area. Thus can result from nozzle wear and other faults resulting in increased flow rates, such as increased pressure and varying travel speed.

Applying agricultural chemicals above label rates is illegal. The consequences of over-application are:

- wasting agricultural chemicals, time and honey
- possible damage to crops (phytotoxicity)
- possibly exceeding maximum residue limits (MRLs)
- extra wear and tear on equipment
- a possibility that the product may not be as effective
- increased risk of off-target damage
- the likelihood that the user will face prosecution.

Under-application of an agricultural chemical occurs when less than the recommended amount of active ingredient is delivered to the target. This can result from blocked nozzles or filters, or when the flow travels faster than the calibrated speed. This problem is difficult to detect with the eye and often goes unnoticed until a major blockage occurs. In some circumstances, it is illegal to apply agricultural chemicals under label rates.

The consequences of under-applications are :

- wasting agricultural chemicals, time and money
- reduced effectiveness of the product
- a greater chance that the pest will develop a resistance to the agricultural chemical
- possible production losses due to pest damage or competition.

Regular calibration of equipment will help you to identify and reduce these problems.

Do not rely on experience to know how far a tank will spray.

Calibration is the only way to check the spraying unit's application rate per area and identify problems in output.

Calibration techniques

Calibrating spray equipment involves four steps. These four steps are applicable to all types of spraying units, including hand-held equipment, boom spraying units, orchard spraying units and even aircraft. The basic principles are discussed in this section. The following steps apply these principles to the major types of equipment;

A. Measuring spraying unit output (L/min)

B. Calculating the area sprayed (m²/min)

C. Calculating the spraying unit's application rate (L/ha)

D. Calculating the amount of formulated product required per spray tank volume

It is important that accurate records are kept of the calibration process. **Before commencing calibration**

It is important to ensure that the spraying unit is operating correctly before taking any measurements with it.

To determine the calibration of spraying units, the user must first check the following and make records where required:

- spraying unit (should be clean and filled with the required volume of water for the exercise)
- label recommendations
- pressure gauge (should be operational)
- pressure regulator setting, if one is fitted
- spray lines and filters (should be free of leaks and blockages)
- nozzle and spraying unit description
- environmental conditions.

A. Measuring spraying unit output (L/min)

The spraying unit output is calculated by collecting the nozzle(s) output, at the operating pressure required, into a container for one minute and measuring the volume collected. The output from all nozzles should be measured. To reduce errors, this procedure should be performed at least three times. Then calculate an average reading.

The measured output of a nozzle is only acceptable if it varies by less than 10% from the manufacturer's specifications, otherwise the nozzle should be replaced.

B. Calculating the area sprayed (m'/min)

To calculate the area sprayed during the calibration, the user needs to have the following two measurements:

- the travel distance. This is the distance travelled by the spraying unit in metres during one minute (normally measured in metres per minute) while actually spraying with water
- the spraying unit's effective swath width in metres. Swath width is the width of spray coverage that is effectively delivered by the nozzle(s) to the target area, i.e. how wide the spraying unit can effectively spray.

The area sprayed in square metres per minute (m^2/min) is calculated using the following formula:

Distance travelled (m/min)	x	Swath width (m)	=	Area sprayed in square metres per minute (m^2/min)

C. Calculating the spraying unit's application rate (L/ha)

This step is used to determine the spraying unit's nozzle output over a given area. For liquids, this is known as the application rate in litres per hectare (L/ha). Users must apply agricultural chemicals at the same application rate that is given on the label.

Spraying unit application rate is calculated by using the data collected in steps A and B.

The following formula is used to calculate the spraying unit's application rate:

Spraying unit output (L/min)	÷	Area sprayed (m^2/min)	x	10000	=	Application rate of the spraying unit (L/ha)

D. Calculating the amount of formulated product required per spray tank volume

This is a critical step in the calibration procedure as it ensures that recommended label rates of agricultural chemical products are applied to target areas, by determining the amount of formulated product to be added to the spray tank.

To calculate the amount of formulated product to add to the desired tank volume, a user needs to have the following information:

- spraying unit application rate (L/ha) (calculated in step C)
- product label rate (L/ha or g/ha or kg/ha)
- tank volume (L).

That information is used in the following formula :

Tank volume (L)	÷	Spraying unit application rate (L/ha)	x	Product label rate	=	Amount of product per tank

General Calibration Procedure

Step A Measuring the spraying unit output (A)

A is the total output from all nozzles in one minute.

A= ______________________________ (L/min)

Step B Calculating the area sprayed (B)

(i) Measure the swath width ___________ (m)

(ii) Measure the speed of the spraying unit ________ (m/min)

(iii) Calculate area sprayed :

B = Speed (m/min) x swath width (m)

B = _______________ (m/min) x _______________ (m)

B = ______________________ (m^2/min)

Step C Calculating the spray application rate (C)

C = A (Spraying unit output) ÷ B (area sprayed) x 10000

C = ____________ (L/min) ÷ ______ (m^2/min) x 10000

Step D Calculating the amount of formulated product required per spray tank volume

(i) Product label rate = ____________________ (L/ha or g/ha or kg/ha)

(ii) Tank volume = __________________ (L)

D = tank volume ÷ **C** (spray application rate) x product label rate

D = ________ (L) ÷ ________ (L/ha) x ________ L/ha or g/ha or kg/ha

D = ____________ (L or g or kg) of product per spray tank volume

Hand-held spraying unit calibration

The general calibration procedure, given above, is suitable for use with all hand-held spraying units. To determine swath width, spray onto a bare, concrete surface and measure the wetted area. Alternatively, use water-sensitive paper.

Application rates for products that are applied by hand-held equipment may be given on the label as either an amount per hectare or an amount per 100 litres of water, Some products specify both. If the label only gives the amount per 100 litres, it may be necessary to change either the flow rate (step A) or work rate (step B), so that the application rate calculated in step D does not exceed the label rate.

Boom spraying unit calibration

The general procedure, given above, can be used with boom spraying units. The main difference is that a boom contains more than one nozzle. It is necessary to measure the output from each nozzle. The spraying unit output (step A) is the sum of all nozzles. The swath width from boom spraying units is commonly the distance between the outer two nozzles, plus one nozzle spacing.

A different method for calibrating boom sprays is given below. This can be used instead of the general procedure.

Orchard spraying unit calibration

(This section adapted from ChemCert Training, developed by Roberto Battaglia)

The calibration of air-assisted spraying units is a two-stage exercise. The air flow is measured first, as the air volume is determined by changing the number of nozzles, nozzle size and also the operating pressure.

The second stage is matching the tractor speed to the available air volume that the spraying unit produces to determine the efficiency of the whole spraying operation.

Often spraying units are operated at a speed which is too fast for the volume of air they produce. Increases in spray volume cannot compensate for inadequate air penetration of the tree canopy.

Air flow calibration for air-assisted spraying units

Step 1 - Air inlet area calculation

1. Measure diameter of air inlet ________ = D (m)
2. Calculate air inlet area using Formula 1 = __________ IA (m^2)

Formula 1 IA = 0.79 x D x D

Step 2 - Air speed measurement

Operate spraying unit fan at recommended revolutions per minute (rpm) (e.g. 540 rpm at PTO) and determine the average airspeed at the inlet (see Figure 9.5). = ________ = S (m/sec).

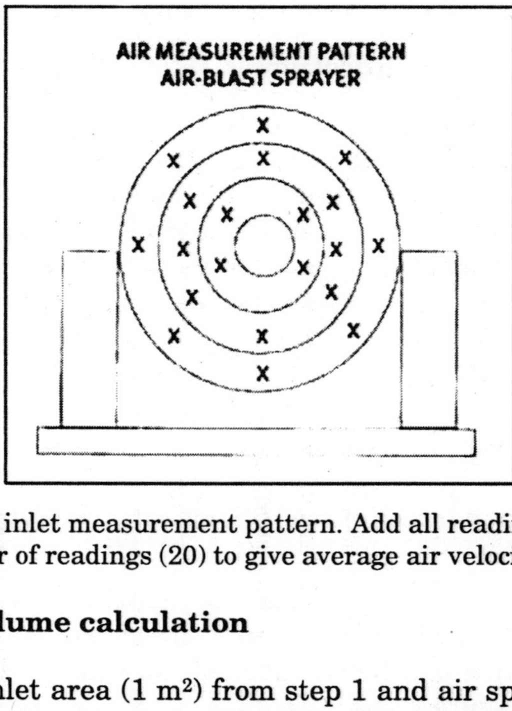

Figure 9.5. Air inlet measurement pattern. Add all readings and divide by number of readings (20) to give average air velocity (m/s)

Step 3 - Air volume calculation

Using air inlet area (1 m^2) from step 1 and air speed (S m/s) from step 2, calculate the air volume (V m^3/h) using Formula 2.

Formula 2 V = E x I x S X 3600

E = entrainment factor - this factor is an estimation of the amount of still air which is drawn into the air stream created by the spraying unit fan (see Figure 9.6).

Open orchard canopy (e.g. apple) - factor of 3 to 3.5 for **E**

Dense orchard canopy (e.g. citrus) - factor of 2 to 2.5 for **E**

Spraying unit calibration sheet

A. Check nozzle—measure total spray output

....

....

....

....

....

Total spray output = ... (mL/min) ÷ 1 000
= .... (L/min)

B. Check actual ground speed

speed (m/min) = distance ...... (m) x 3.6 ÷ time taken .... (min)
= .... km/h

Measure effective spray width

.... metres (m)

C. Calculate spray application rate (litres per sprayed ha) at a measured speed

Spray application rate = spraying unit output (L/min) x 600 / spray width (m) x speed (km/h) = ... L/min x 600 ÷ .... m ÷ .... km/h = ... L/ha

or

Calculate speed to travel to give a desired spray rate (L/ha)

Speed (km/h) = spraying unit output ....(L/min) x 600 ÷ effective spray width .... (m) ÷ desired application rate .... (L/ha) = ... km/h

Time to cover 100 m = 360 ÷ speed .... (km/h) = ... seconds

or

Calculate spraying unit output to give a desired spray rate (L/ha) at a given speed

Spray output required =

Desired application rate .... (L/ha) x effective spray width .... (m) x speed ....(km/h) ÷ 600 = ... L/min

Required output per nozzle =

Total spraying unit output .... (L/min) ÷ .... number of nozzles = ... L/min/nozzle x 1000 = ... mL/min/nozzle

To calculate spray mixes

Total amount of formulated product required (L, mL, kg, or g)	or	**Total amount of formulated product required (L, mL, kg, or g)**
= Total area to spray .... (ha) x Product label rate .... (......./ha) = ... (ha)		= Product label spray concemtration .... (....../100L) x Product spray quantity .... (L) ÷ 100 = ... (.........)

Total quantity of spray required = total area sprayed .... (ha) x spray application rate .... (L/ha)

= L

Total number of tanks of spray = total quantity of spray required .... (L) ÷ tank size .... (L)

= .... **tanks**

Amount of formulated product for each *full* tank =

Volume of water in tank ..... (L) x product label rate .. (.....) ÷ spray application rate .... (L/ha)

= .... (.....)

Amount of formulated product for each part tank =

Volume of water in tank .... (L) x product label rate .... (......) ÷ spray application rate .... (L/ha)

= .... (.....)

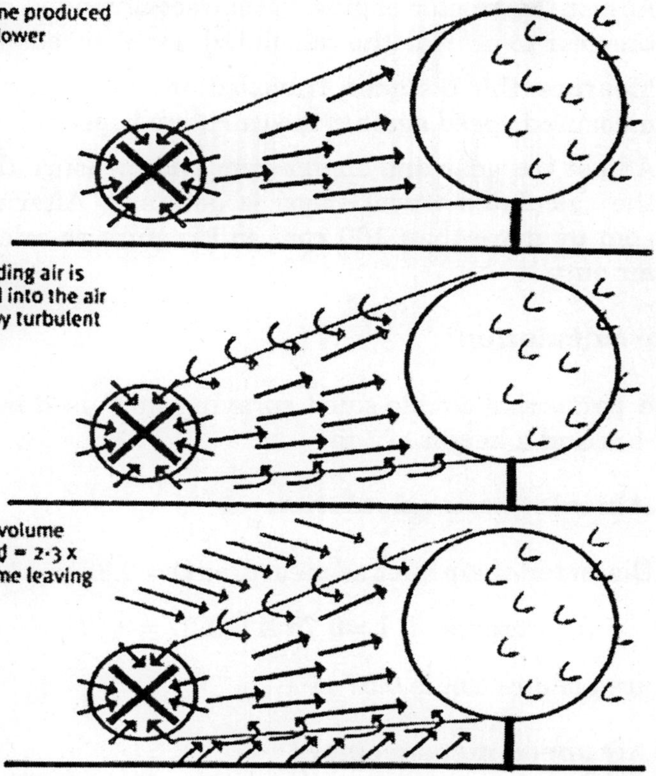

Figure 9.6. The entrainment factor in orchard tree spraying

Step 4 - Matching tractor speed to air volume

1. Measure width of tree canopy _______ = W (m)
2. Measure height of tree canopy ______ = H (m)

For single-sided spray

3- Calculate travel speed of tractor using Formula 3 ______= T (km/h)

Formula 3 T = V ÷ W ÷ H 1000

For double-sided spraying unit

4. Calculate travel speed of tractor using Formula 4 _______=T (km/h)

Formula 4 T = V ÷ 2 ÷ H ÷ 1000

1. Adjust the tractor engine to the recommended rpm and select the gear to provide the calculated travel speed of the tractor.
2. Measure the distance travelled in one minute. Check the measured speed against the calculated speed.
3. Adjust the gear and engine rpm and measure the speed until the calculated travel speed is obtained. Altering the engine rpm by more than 100 rpm will require re-calculation of the air output.

Example calculation

For a particular double-sided spraying unit used on trees with a width of 4 m and a height of 5 m:

Step 1 - Air inlet area calculation

1. Diameter of air inlet as measured D = 0.85 m

Formula 1 $I = 0.79 \times D \times D = 0.57\ m^2$

The area of air inlet is 0.57 m^2.

Step 2 - Air speed measurement

Air velocity measurements:

12, 14,17,13,15,15,17,14,12,18,16,18,14,13,16,17,14,15,16,14

Average airspeed at the inlet = 300÷20 = 15 m/s.

Step 3 - Air volume calculation

Using air inlet area (I) from step 1 and air speed (S) from step 2, calculate air volume using Formula 2.

Formula 2 $V = 2 \times 0.57 \times 15 \times 3600 = 61560\ m^3/h$

Step 4 - Matching tractor speed to air volume

1. Width of tree canopy W = 4 m
2. Height of tree canopy H = 5 m
3. Calculate travel speed of tractor using Formula 4

Formula 4 $T = 61560 \div 2 \div 4 \div 5 \div 1000 = 1.5 km/h$

The correct spraying speed in this situation is 1.5 km/h.

Your calculation

Step 1 - Air inlet area calculation

1. Diameter of air inlet D = ________ m

Formula 1 I = 0.79 x D x D (m^2) = 0.79 x ___ x ___ = ______ m^2

Step 2 -Air speed measurement

Air velocity measurements: __________________

Average airspeed at the inlet S = ___________m/sec.

Step 3 - Air volume calculation

Using the air inlet area (I) from step 1 and the air speed (S) from step 2, calculate the air volume using Formula 2.

Formula 2 V = E x I x S x 3600 = ______ x _____ x 3600 = _____m^3/h

Step 4 - Matching tractor speed to air volume

1. Width of tree canopy W =_____ m
2. Height of tree canopy H = _____m
3. Calculate travel speed of tractor using Formula 3 or 4

Formula 4 T = V ÷ 2 ÷W ÷ H ÷ 1000 = _____÷ 2 ÷
____ ÷ _____ ÷ 1000 = _____ km/h

Application volume (L/ha)

Most orchard chemicals are applied as a proportion of the total volume of liquid applied. As nozzles wear and their output volume increases, more agricultural chemical will be applied to the orchard. Worn nozzles also produce larger droplets, which result in reduced coverage and increased losses of agricultural chemical to the environment, particularly the ground. Calibration improves spray application efficiency by enabling appropriate amounts of agricultural chemical to be added to a spray tank. Regularly checking the nozzle flow rate and replacing worn nozzles also enhances spray efficiency, as the nozzles are more likely to produce the appropriate range of droplet sizes.

Step 1 - Measurement of tractor speed

For an air-assisted spray, the tractor speed will be determined by the air flow calibration. For other types of spraying units, operate the tractor in the gear and rpm that will be used when spraying. Use one of the following methods to calculate the tractor speed that will be used when spraying.

Peg out a 50 m or 100 m track in the orchard and measure the time taken for the tractor to cover the track (seconds).

Calculate tractor speed

Divide the track length (m) by the time taken (s), and multiply by 3.6 to give the speed (km/h).

Tractor speed (km/h) = distance (m) x 3.6 ÷ time taken (s)

Step 2 - Nozzle flow calibration

Operate the pump at a normal rpm and operate the nozzles at a normal spraying pressure. Check the pressure at the nozzle and compare it to the main gauge. Clip-on plastic tubes provide the simplest method for directing spray volume into measuring jugs.

Record individual nozzle flow for each position on the spraying unit for a standard time (such as 1 minute).

Record the output for each nozzle:

1	L/min	8	L/min
2	L/min	9	L/min
3	L/min	10	L/min
4	L/min	11	L/min
5	L/min	12	L/min
6	L/min	13	L/min
7	L/min	14	L/min
Total output			

Add the flow rate for each nozzle to give the total output for the spraying unit. Replace any nozzle that varies by more than 10% from the manufacturer's specifications.

Step 3 - Calculate application volume L/ha

Using a calculator:

L/ha = total output (L/min) x 600 ÷ tractor speed (km/h)
– effective width (m).

For a double-sided spraying unit, effective width = row width For a single-sided spraying unit, effective width = row width ÷ 2

Example calculation using method 1

Step 1 - Measurement of tractor speed

For a double-sided spraying unit on trees with tree spacing of 6 m and row spacing of 6 m (effective width - 6 m).

Distance measured (metres) = 100 m

Time taken (seconds) = 200 S

Calculate tractor speed:

Tractor speed (km/h) =

distance (m) x 3.6 ÷ time taken (s)

= 100 x 3.6 ÷ 200

= 1.8 km/h

Step 2 - Nozzle flow rate

1	2.1	L/min	8	2.1	L/min
2	2.1	L/min	9	2.2	L/min
3	2.1	L/min	10	2.3	L/min
4	2.2	L/min	11	2.2	L/min
5	1.6	L/min	12	1.8	L/min
6	1.7	L/min	13	1.6	L/min
7	1.8	L/min	14	1.9	L/min
Total output				**27.7**	**L/min**

Step 3 - Application volume L/ha

Application volume:

(L/ha) = total output (L/min) x 600 ÷ tractor speed (km/h)
÷ effective width (m).

Using a calculator:

(L/ha) = total output (L/min) x 600 ÷ tractor speed (km/h)
effective width (m).

=27.7 x 600 ÷ 1.8 ÷ 6

= 1539 L/ha

Your calculation using method 1

Step 1 - Measurement of tractor speed

Calculate tractor speed:

Tractor speed (km/h) = distance (m) x 3.6

= time taken (s)

= _____ x 3.6 ÷ ______

= _____ km/h

Step 2 - Nozzle flow rate

1	L/min	8	L/min
2	L/min	9	L/min
3	L/min	10	L/min
4	L/min	11	L/min
5	L/min	12	L/min
6	L/min	13	L/min
7	L/min	14	L/min
Total output			

Step 3 - Application volume L/ha

Application volume :

L/ha = total output (L/min) x 600 ÷ tractor speed (km/h)
÷ effective width (m)

= _________ L/min x 600 ÷ _______ km/h ÷ _______ m

= _______ L/ha

AGRICULTURAL CHEMICAL DRIFT MANAGEMENT

Introduction

This section discusses procedures and techniques to reduce the potential for spray drift from spray applicators.

The user must read and understand the agricultural chemical product label before spraying and follow the label instructions during spraying.

Definition of spray drift

Spray drift is the movement of an agricultural or other chemical (as droplets in air) outside the intended target area. Spray drift is also defined as off-target movement of an agricultural chemical that has the potential to cause injury or damage to humans, plants, animals, environment or property.

Spray drift does not include off-target move rent of an agricultural chemical due to post-spray volatilisation (evaporation) or movement in water, soil or organisms.

The causes of spray drift are one, or a combination of, the following:

- spraying in unsuitable weather conditions
- using spray equipment inappropriately (i.e. when the use is inconsistent with the manufacturer's instructions or training guidelines)
- using an unsuitable (e.g. unregistered or unapproved) agricultural chemical formulation for a particular use or in a particular area
- using a droplet size that is too small

Property management planning

The management of spray drift should be actively considered as part of property management planning and farm or enterprise development.

Spray drift management starts when a farm or enterprise is planned or developed. During this phase of land development, users can help to identify areas where agricultural chemical application may conflict with neighbouring laid use. An essential part of spray drift management is adequate property management planning to avoid potential conflict between different land uses. Farm plans should include adequate separation between cropped areas, requiring agricultural chemical spraying, and ether land, which may be affected by those agricultural chemicals.

Preparation of a property management plan

Landholders, in consultation with agricultural chemical users or operators, should develop a comprehensive property management plan before planting any crop. The plan should take into account requirements associated with applying agricultural chemicals.

Development of awareness zones

Landholders, in consultation with agricultural chemical users or operators, should:

- establish a spray drift awareness zone around the field to be sprayed
- identify all areas within the awareness zone that may be affected by spray drift
- produce an awareness zone chart identifying areas at risk of spray drift
- make an updated awareness zone chart available to the agricultural chemical user or operator before any spraying operation
- prepare a plan for all routine spraying operations. This will include instructions on awareness zone location and drift-reduction buffer zones, as well as safety measures specific to the site (see Figure 10.1).

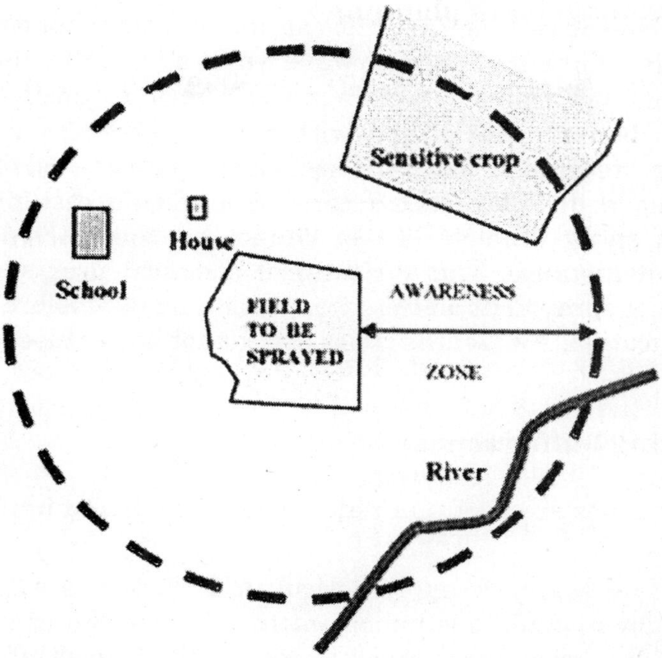

Figure 10.1. Awareness zone establishment and sensitive area identification

The agricultural chemical user, together with the landowner, should be responsible for establishing a spray drift awareness zone around the property. Under most circumstances, the awareness zone for ground spraying could extend from approximately 100 m to 1 km from the paddock to be sprayed and up to 5 km from the paddock for aerial spraying. The width of the awareness zone may be extended if there are sensitive areas nearby, or if the area is subject to local meteorological effects, such as cold air drainage in a valley (katabatic wind flows - see Section 8).

Spray operators should develop their own appropriate spray drift assessment zones. The zone should be used to help the survey of areas or buildings outside a field to be sprayed that may be potentially sensitive to spray drift. Areas or buildings that could be at risk include schools, houses, wetland areas, traveling stock routes and organic farms.

Important Note

As stated above, it is important to survey the awareness zone for areas sensitive to spray drift. Ideally, these areas should be identified before planting a crop that may require spray application as well as before any spraying operation begins. Areas or buildings that are potentially sensitive to spray drift can include neighbouring sensitive crops, native flora and fauna, waterways and wetlands, bees, off-target plants and animals, buildings and homes. If sensitive areas are present or become established within the awareness zone, details should be recorded and if necessary, expert advice sought. Establishing a drift awareness zone at a certain distance does not mean that spray droplets can or cannot be transported in the air beyond that distance. The agricultural chemical user should assess the impact of spray drift on the awareness zone just before and during spray application. Awareness zones should not be confused with buffer zones.

Establishing buffer zones

Buffer zones are useful in reducing the downwind impact of spray drift.

If it is necessary to apply agricultural chemicals when sensitive areas are down wind, it is important to establish buffer zones on the downwind side of sprayed areas to reduce the impact of spray drift. Buffer zones are usually located on the downwind side of a sprayed area and are used to protect an area at risk of off-target spray movement. Some agricultural chemical labels give downwind buffer distances for specific crop or product situations. The distance required for a buffer zone will depend upon factors such as the type of buffer zone, weather and application method.

In-crop buffer strips

- Spray operators should consider spraying only the upwind section of a field, so that the downwind unsprayed section will catch any spray drift. This is called field splitting (see Figure 10.2).
- Spray operators should leave the last row or swatch unsprayed on the downwind boundary of fields next to an area that is at risk.

Unsprayed areas should be sprayed when the wind is blowing away from the areas that are at risk. This also has the advantage of acting as a comparison check for the agricultural chemical's performance. Buffer strips reduce spray drift by adsorbing particles onto the vegetation surface as the air flows over the buffer area. Subsequent follow-up sprays should take into account the possibility that the initial unsprayed area may have caught spray drift. So, the unsprayed area may hold chemical deposits that exceed maximum residue limits (MRLs).

Figure 10.2. An example of field splitting

Spray operators should not assume that a butter zone is the only method necessary to avoid spray drift. Users should follow responsible spraying practices, whether or not a buffer zone is available, by using correctly calibrated equipment in good condition, spraying in appropriate weather conditions and spraying strictly in accordance with product label instructions. Spray operators should still observe good spraying practices when using a buffer zone.

Vegetative barriers

Plant and maintain buffer vegetation barriers on downwind edges of fields and properties as well as next to areas that are at risk from spray drift.

A vegetative barrier is usually a tree or shrub line located on the downwind side of a sprayed area. The vegetative barrier s used to protect an area that is at risk from spray drift. Vegetation is sometimes planted deliberately to filter spray drift from the environment. However, planting vegetative buffers should be regarded as additional to the main methods of drift reduction already discussed. Vegetation barriers reduce spray drift by filtering the air as it flows through the porous (absorbent) barrier of leaves.

Fallow buffer strips

Where an in-crop or vegetation buffer cannot be used to lessen the impact of spray drift, reduce spray drift landing off-target by establishing a downwind, open-fallow buffer zone (see Figure 10.3).

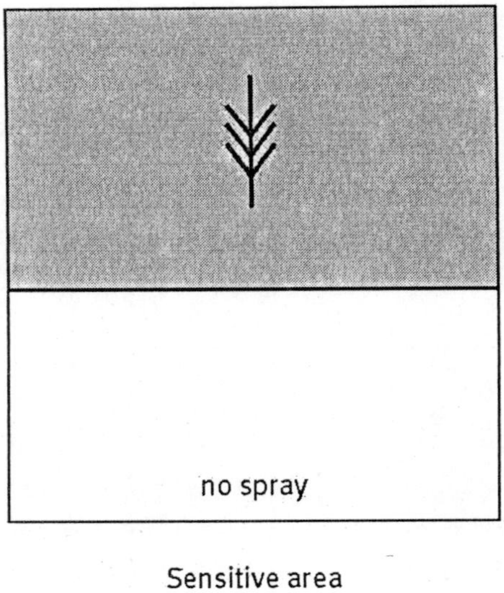

Figure 10.3. A no-spray area between the area to be sprayed and the sensitive area

Larger buffer zones may be required where in-crop or buffer vegetation cannot be used to catch spray drift. Fallow buffer zones rely mainly on distance and can vary considerably, depending upon specific application situations and label requirements. For example, 200 m is the distance required for the ground application of insecticides, containing endosulfan, to cotton.

Communication

Communicating with neighbours and stakeholders about proposed spraying activities will help to develop cooperative spray management strategies as well as helping to avoid future conflicts.

Many spray drift incidents can be avoided, or their impact reduced, if neighbours, contractors and sometimes the local community are advised and consulted before application. In some areas, local chemical liaison committees exist to promote close working relationships with growers, agricultural chemical users and concerned community members.

Who, when and how?

The agricultural chemical user should communicate with neighbouring properties and other stakeholders about anticipated spraying schedules and activities before the season and before spray application, particularly when that information has been asked for.

All parties should try to agree on the method of communication. Contact by phone may be sufficient in some situations, while other situations may require written advice.

Regularly communicate with local agricultural chemical liaison committees, consultants and applicators. Detailed communication may be needed for specific neighbours who live in an area at risk from spray drift.

Strategies for reducing chemical risk

Consider alternatives

Think about other methods of pest control before agricultural chemical application (e.g. cultural, mechanical or biological control). If chemical control is selected, choose registered products that are least harmful to off-target organisms and the environment.

Understand the product

The agricultural chemical user must read, understand and follow the label of the product to be used. The user should also have access to and read a current material safety data sheet (MSDS).

Maintain equipment

- Maintain and operate spray equipment according to the manufacturer's instructions.
- Regularly check all spray equipment for wear, damage, leaks and specification (nozzle type etc.).
- Regularly calibrate and check the perform ante of all spray equipment to ensure the spray operator applies the correct application rate of material.
- Keep calibration records.
- Decontaminate spray equipment after use, following any label instructions.

Check weather

Weather conditions should be within acceptable limits for safely and effectively applying agricultural chemicals.

The weather plays an important role in controlling agricultural chemicals applied as sprays. The spray operator should check that weather conditions are within acceptable limits before spraying. It is essential that the person spraying is aware of the effects of wind speed, wind direction, temperature, humidity and, most importantly, atmospheric stability on spray drift. Inexpensive, hand-held anemometers and psychrometers are available to monitor these conditions. Larger enterprises regularly involved in the application of agricultural chemicals should purchase meteorological stat on data loggers. Spray operators should discuss weather conditions with the landholder who has contracted them to carry out the spraying, as the landholder will have more knowledge of local conditions and changing weather patterns. The spray operator and landholder should together make the decision whether to spray based on the weather.

Wind speed and direction

Measure wind direction and wind speed before and during spray application. This can be done using an anemometer (see Figure 10.4). Record the information in a record logbook.

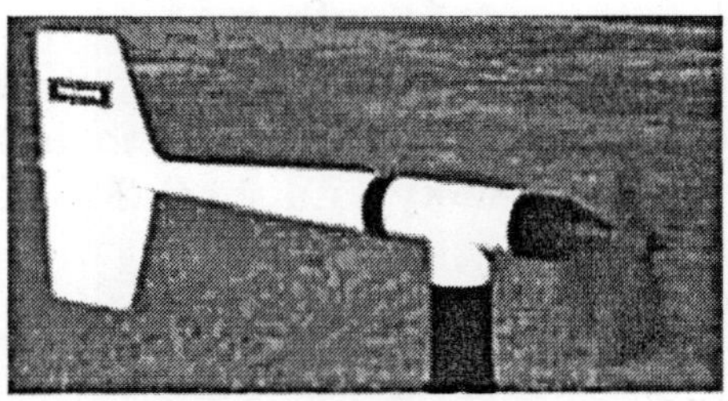

Figure 10.4. An anemometer

- Apply sprays when the wind direction is away from areas sensitive to the agricultural chemical application. (See Figure 10.5)

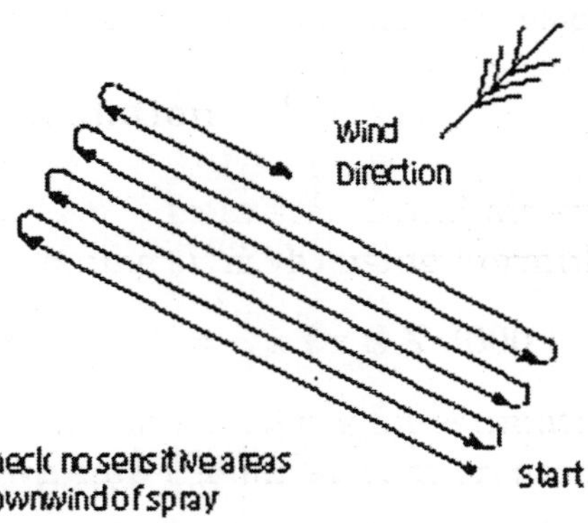

Figure 10.5. Ideal direction of travel when spraying relative to wind direction

- Spraying should not take place if tie wind is light and variable in strength or direction.
- Spray operators should be alert to changes in wind direction and alter or cancel a spray program as necessary.
- Wind speed should be between about 3 km/h and 15 km/h for most spraying operations.
- Spraying should, where possible, be carried out with a cross wind, working into the wind towards the unsprayed area.

The most commonly cited wind speed limit for general purpose spraying is 15 km/h, 8 knots or 4 m/s. Above about 15 km/h, there may be too much airborne movement of medium and large droplets downwind from the intended target area. In general terms, in stronger winds, droplets will travel further. Smaller droplets will also travel further, as their fall speed is much slower. In addition, turbulence and atmospheric stability heavily influence the movement of small droplets.

The spray operator will need to carefully consider the greater risk of spraying in winds exceeding about 15 km/h and will need to adopt specific strategies to avoid spray drift impacting in an off-target area. Under these circumstances, the spray operator should establish a wide buffer zone outside the downwind boundary of the sprayed area.

Temperature and humidity

- Spraying should ideally take place when temperatures are most favourable (in a 24-hour cycle).
- Spraying of water-based sprays should not take place under conditions of high temperature and low humidity, for example, when the wet bulb depression (a measure of evaporation potential) is greater than about 10°C.

Whenever possible, avoid spraying in high air temperatures. Water-based sprays evaporate, particularly when the air temperature is high and the relative humidity is low. Evaporation reduces the droplet size of water-based sprays as the droplets move through the air. Initial droplet size may be increased to make up for this. High temperatures can also mean the onset of unstable atmospheres that may stop droplets reaching the intended target area.

Wet bulb depression is defined as the difference in wet bulb and dry bulb temperatures, as measured by a hand-held whirling psychrometer (see Figure 10.6). High temperatures (e.g. greater than

30^0C) combined with low relative humidity (e.g. less than 40%, i.e. a wet bulb depression greater than 10^0C) can increase the rate of evaporation of water-based sprays and may also increase spray drift.

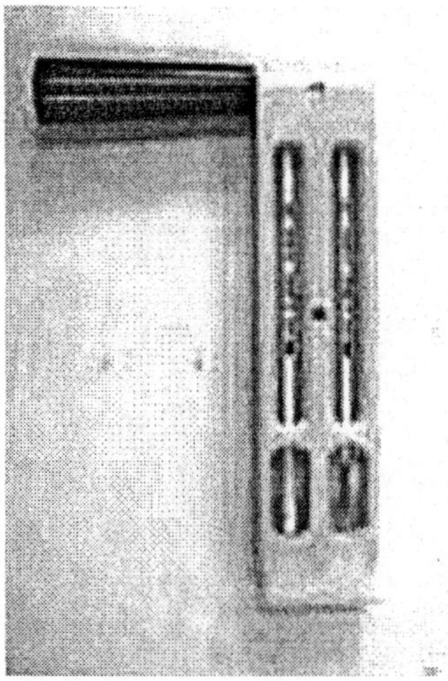

Figure 10.6. Whirling psychrometer

Atmospheric stability

The term stability describes the vertical movement of air in the atmosphere. If atmospheric conditions are unstable, the dispersion (or spread) of spray upwards may be high, increasing the amount of spray that enters the atmosphere (see Figure 10.7). In contrast, the dispersion rate of droplets may be low in stable atmospheric conditions, leading to high amounts of spray landing off-target at ground level.

If the sky is clear at night, the ground can lose heat rapidly in the dry atmosphere and cool air layers form close to the soil surface. Under these conditions, air close to the ground is cooler than the air above. Since this is opposite to the normal condition of the atmospheretemperature decreasing with height-the condition is called a surface temperature inversion.

Temperature inversions tend to suppress the vertical movement of air and, therefore, present a barrier to the transport of small droplets to the crop canopy. Inversions usually form under very low wind speed conditions. Avoid spraying under such circumstances, since small droplets are capable of remaining airborne for long periods within the inversion layer and can cause severe damage several kilometres away from where spraying took place.

Spraying should therefore, ideally take place in neutral atmospheric conditions (see Figure 10.7). The stability of the atmosphere can be tested using smoke, or by driving a vehicle along a dusty track. If smoke or dust collects within a thin layer, it can indicate the presence of a surface temperature inversion.

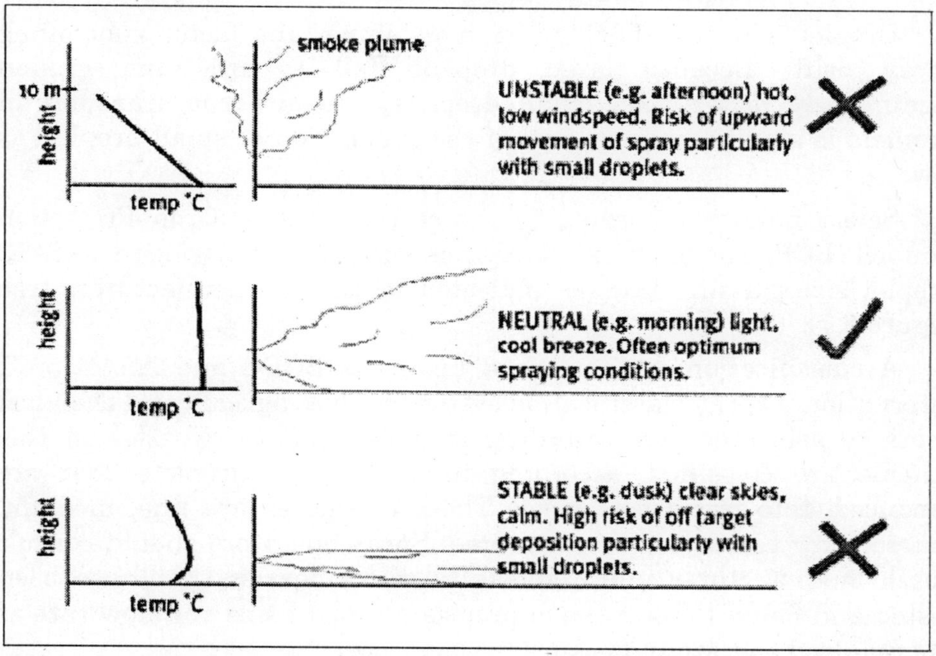

Figure 10.7. Basic guide to air stability

Spray application technology

Operate spray application systems so that they minimise the potential for spray drift. In summary:

1. Spraying should ideally take place when atmospheric conditions are neutral.
2. Do not spray during highly unstable conditions.
3. Spraying should not take place when surface temperature inversion conditions exist.
4. Where appropriate, use a smoke-generating device to determine atmospheric stability.

The most appropriate spray application equipment for a particular task will deliver droplets of the appropriate size in a way that increases their chances of landing on the target. Particular types of spray application equipment are designed for specific purposes. For example, application equipment designed for broad-acre crops is unlikely to be suitable for use in orchards. Spray application equipment should be properly maintained, calibrated and operated to maximise efficiency and avoid excessive spray drift.

Droplet size

Droplet size is probably the most important factor controlling spray drift. Because large droplets fall towards the ground significantly faster than small droplets, the airborne transport of droplets is significantly reduced if the production of small droplets is low.

Select nozzles according to the correct British Crop Protection Council (BCPC) or American Society of Agricultural Engineers (ASAE) droplet size classification, as indicated by nozzle manufacturers (see Figure 8.7).

A classification system based upon the BCPC and ASAE 5572 international spray classification system enables agricultural chemical users to select nozzles according to spray quality. Nozzles in this scheme are classified, according to the size of droplets that are produced, into broad categories. These categories are fine, medium, coarse, very coarse and extra coarse. Spray operators should consult nozzle manufacturer's data sheets to select nozzles that have been tested and found to have the appropriate droplet size and flow rate at the required operating pressure.

Release height

- Release height (h) of the spray (see Figure 10.8) should be as low as possible, and consistent with nozzle specifications, coverage requirements and any label directions.
- Boom height should not exceed optimum heights specified by the nozzle manufacturer.

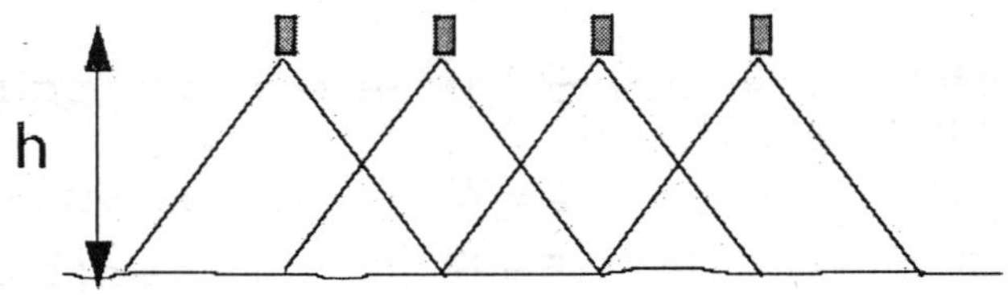

Figure 10.8. Release height

Release height of the spray is an important factor influencing spray drift levels. The higher the droplets are released, the greater the potential for drift. Adequate boom stabilisation is essential, especially on uneven ground. Boom height may be lowered to produce less spray drift, although adjusting nozzle number, type and orientation is usually required to maintain an even spray pattern. The use of wide-angle, flat fan nozzles (e.g. 110°) usually permits lower boom heights to be used.

Spray pressure

- Spray pressure should be as low as possible, and consistent with nozzle specifications and coverage requirements.

When the pressure of a ground rig spray boom is increased, most hydraulic nozzles produce smaller droplets. To reduce drift potential, low pressures should be used. Many nozzle manufacturers now provide hydraulic nozzle tips that can be operated as low as 1 bar (14 psi). Control spray volume by changing nozzles not by changing pressure (i.e. to increase flow rates select nozzles with larger orifices).

Hand-held application - field and row crops

- Spray operators should keep the movement and height of the spray lance at a minimum.
- Spray nozzles should be held at a constant height that will produce an even spray pattern. It should be consistent with nozzle specifications and coverage requirements.

Orchard spraying

- Spray operators should ensure that air blast spraying units are configured to directly spray into the crop canopy.
- Air blast spraying units (see Figure 10.9) should not be used to spray the downwind edge of an orchard (downwind side of the outer row) when areas at risk within the awareness zone are located downwind of the field edge.

Air blast spraying units direct high-speed air into the crop canopy. Spray operators should ensure that spray droplets are contained within a tree canopy and not directly sprayed into the air above a canopy. Where appropriate, use air deflectors so that the air produced by an air blast spraying unit is directed into the canopy and not lost between the rows.

Identify downwind areas that are at risk within the awareness zone during spraying. Take steps to reduce drift by using:

Figure 10.9. Air blast spraying unit

- existing rows of trees adjacent to the area at risk of drift as unsprayed buffers
- equipment that directs the air out one side of the spraying unit and spraying the other row or rows from one direction only (away from the area at risk of drift).

Record keeping

The spray operator must maintain up-to-date records of all agricultural chemical spraying operations carried out.

The spray operator should produce a spray report after every spray application. The report should include the date, time, area sprayed, amount and name of agricultural chemicals applied, recorded application rates, spray operator(s) involved, equipment used, nozzle type, settings (spray pressure, blade angle or unit rpm where appropriate) and meteorological conditions (wind speed and direction, temperature and humidity).

Record keeping is a legal requirement for commercial ground distribution (application) of agricultural chemicals, involving licensed operators, in Queensland. Licensed operators, carrying out such commercial activities on behalf of licensed ground distribution businesses, must ensure that their records of all spray applications conform to legal requirements. Records of commercial activities need to be kept for a minimum time period (presently in Queensland, this time is for at least two years). In addition, government enforcement officers may monitor these records. Despite the legal requirement of at least two years, it is prudent to maintain records for considerably longer (e.g. 10 years or more) in case a neighbour or other party seeks compensation in a court action. Production of records may be useful to demonstrate, for example, that the spray operator, at the given time, correctly applied the agricultural chemical product to the target crop or situation.

- Record the following information:
  - awareness zone chart - calibration report
  - spray report, including weather observations
  - agricultural chemical listing - a register of all agricultural chemicals used.
- The agricultural chemical user should hold copies of material safety data sheets (MSDSs) for all products used on the property or by the business.
- Agricultural chemical users should ensure that material safety data sheets are readily available and consulted prior to application.
- Agricultural chemical users should retain records in accordance with relevant legislative requirements
- MSDSs are available from the supplier or manufacturer of the farm chemical or from Infopest.

Summary of drift management

Large droplets and buffer zones can reduce spray drift. However, such strategies should always be used together with other valuable techniques. The use of single strategies may not be enough to have a significant effect on spray drift management.

How spray operators can reduce spray drift:

1. Identify all areas, around an area to be sprayed, which could be at risk from spray drift damage.
2. Communicate on a regular basis with neighbours regarding proposed spray schedules and activities.
3. Keep copies of relevant material safety data sheets (MSDSs).
4. Read, understand and follow label instructions before and during spraying.
5. Observe and record wind direction, wind speed, temperature and humidity before and during spray application, including details of changes in conditions.
6. Avoid spraying when wind is blowing towards areas at risk from spray drift.

7. Do not spray if the wind is light and variable in strength or direction.
8. Spray water-based sprays when temperatures are most favourable (in a 24-hour cycle).
9. Do not spray water-based sprays under conditions of high temperature and low humidity.
10. Spray when atmospheric conditions are neutral.
11. Do not spray during highly unstable condition.
12. Do not spray when surface temperature inversion exists.
13. Where appropriate, spray on the upwind section of a field, so that the unsprayed downwind section is used to catch any spray drift (field splitting).
14. Carry out spraying, where possible, with a cross wind and move up wind.
15. Use smoke-generating devices to monitor changes in wind direction and stability.
16. Correctly calibrate all spray equipment.
17. Be prepared to stop a spraying operation that has already begun if there are changes in weather conditions that are not suitable for continuing the spraying.

References

Primary Industries Standing Committee, 2002, Spray drift management: principles, strategies and supporting information, CSI RO, Melbourne.

THE LAW AND AGRICULTURAL CHEMICALS

Both the Commonwealth Government and the governments of the states and territories have legislation regulating agricultural chemicals. The Commonwealth Government regulates their sale and supply and the states and territories regulate their control of use.

The role of the Commonwealth Government

The National Registration Scheme acts as the starting point. The Australian Pesticides and Veterinary Medicines Authority (APVMA) regulates the manufacture, distribution and supply of products from registration up to, and including, the point of retail sale. Registration is granted to a new agricultural chemical prod pct following a rigorous scientific evaluation that considers factors such as product efficacy and human, animal and environmental safety, as well as the potential impact on trade. This process includes a degree of public consultation.

The manufacturer and potential marketer determine the protocols under which agricultural chemicals may effectively be used. The manufacturer (registrant) submits a registration proposal to the APVMA for consideration. If the proposal is deemed to be satisfactory, an agricultural chemical product is registered by the APVMA under the Agvet Code, which is a Schedule to the Agricultural and Veterinary Chemicals Code Act 1994 (Commonwealth). The Agvet Code is recognised as Queensland law and referred to as the Agvet Code of Queensland by virtue of the Agricultural and Veterinary Chemicals (Queensland) Act 1994.

An integral part of registration is the requirement that precise instructions for the use of products be included as part of the approved product label, and the material accompanying the product, when sold. Labels must contain information such as the approved rate(s) of use, dose rate(s), route(s) of administration, concentration(s) of active constituents, safety precautions and a statement of the withholding period(s).

Under the Agvet Code certain products are declared to be 'restricted chemical products' because of the increased risk associated with their use. Supply and use of such products are restricted to authorised persons. An example in 2005 of a restricted chemical product is the insecticide endosulfan.

The role of the states and territories

The states and territories supplement the Commonwealth registration requirements by adequate and effective control-of-use legislation. Each state and territory has its own control-of-use legislation covering not only agricultural chemical products but also veterinary chemical products.

Queensland legislation

The Queensland legislation is summa riled as follows:

Agricultural Chemicals Distribution Control Act 1966 (ACDC Act) and Regulation

The ACDC Act provides for an Agricultural Chemicals Distribution Control Board with power to grant, refuse, cancel or suspend the following licences:

- **pilot chemical rating licences** issued to agricultural pilots distributing (applying) agricultural chemicals from aircraft (aerial distribution)
- **aerial distribution contractor licences** issued to the aerial businesses directing or authorising aircraft to be used for aerial distribution of agricultural chemicals, or ground equipment to be used for ground distribution of herbicides
- **commercial operator licences** issued to persons distributing herbicides from ground equipment on land that the persons or their near relatives do not own or occupy (ground distribution)
- **ground distribution contractor licences** issued to the ground distribution businesses directing or authorising ground distribution to be used for ground distribution.

The ACDC Act and the Agricultural Chemicals Distribution Control Regulation 1998 (ACDC Regulation) regulate the distribution (spraying, spreading and dispersing) activities of licence holders. An important regulation is that licensed operators must not carry out spraying with equipment or under meteorological conditions that might reasonably be expected to cause damage to off-target crops or livestock.

The Act requires ground or aerial distribution contractors to keep detailed records of each and every spraying operation they direct or authorise to be carried out by licensed operators.

Provision is made for the declaration of hazardous areas in response to some agricultural chemicals having a history of causing damage to off-target crops. Under the restrictions that apply in hazardous areas, use of certain agricultural chemicals are restricted or may be prohibited in the declared area. At time of printing (2005) three hazardous areas have been declared in specific geographic locations.

The Department of Primary Industries and Fisheries administers the ACDC Act and ACDC Regulation.

Chemical Usage (Agricultural and Veterinary)

Control Act 1988 (Chemical Usage Act) and Regulation

The Chemical Usage Act, among other things, confines the use of agricultural chemicals to products registered or approved with the APVMA. Approved products must be used according to label instructions or approved permits, except where the Act clearly sets out the legal ways an agricultural chemical may be used contrary to label instructions. It sets out particular situations where unregistered agricultural chemicals may or may not be used.

This act provides for a number of control mechanisms when MRLs of agricultural chemicals are exceeded in agricultural produce and other prescribed substances.

The Chemical Usage Act also provides for the prohibition (banning), in the public interest, of particular agricultural chemicals. Examples currently banned (2005) include the persistent organo-chlorine insecticides such as those containing DDT, dieldrin, chlordane or heptachor.

An important provision of the Chemical Usage (Agricultural and Veterinary) Control Regulation 1999 (the Regulation) sets out the authorised persons who may be permitted to use declared restricted chemical products such as endosulfan and also the qualifications that are required for these persons to be so authorised.

The Department of Primary Industries and Fisheries administers the Chemical Usage Act and Regulation.

Workplace Health and Safety Act 1995 (WHS Act) and Regulation

The WHS Act influences workplace practices at every level of activity. It has the prime objective of ensuring everyone's health and safety, not just workers and employers, but also members of the public who may be affected by the activities at the workplace. The object of the WHS Act is to prevent people being killed, injured or contracting an illness because of a workplace or a workplace activity. The WHS Act requires those with an obligation for health and safety (e.g. employers, the self-employed, workers, labour hire contractors etc.) to follow a risk management approach, that is:

- identify hazards
- assess risks that may result because of the hazards
- decide on control measures to prevent or minimise the level of the risks
- implement control measures
- monitor and review the effectiveness of measures.

This means that people need to think about their actions in the workplace and carry out their duties in a responsible way without putting anyone at risk. This is called showing a 'Duty of Care' and negligence in this area is regarded as a serious failing and may lead to prosecution.

Duty of Care responsibilities apply equally to both employer (e.g. minimising over-spray of agricultural chemicals, ensuring that agricultural chemicals are used safely, providing adequate training and PPE and washing facilities) and employees (attending courses, following instructions for mixing and applying agricultural chemicals safely, wearing PPE).

Particular regulations may apply to certain types of agricultural chemicals in addition to the normal duty of care expectation, e.g. organophosphate pesticides are a scheduled hazardous substance under Part 13 of the Workplace Health and Safety Regulation 1997 and those mixing or using these products require regular blood tests.

The main thing to remember is that you must obtain a material safety data sheet for each agricultural chemical that you use and you must follow the label directions.

For a summary of the wH5 Act 1995 go to: http://www.dir.qld.gov.au/ workplace/law/legislation/

For information on risk management process, and management of hazardous substance for employers and workers, go to:

http://www.dir.gld.gov.au/workplace/subjects/

The Workplace Health and Safety Division of the Department of Industrial Relations administers the WHS Act and Regulation.

Dangerous Goods Safety Management Act 2001 (DGSM Act) and Regulation

Some agricultural chemicals used on farms will be classified as Class 6 Dangerous Goods under the Dangerous Goods Safety / Management Act 2001 (DGSM Act). They may also have a Class 3 'flammable liquids' subsidiary risk to consider.

The overall objective of the DGSM Act is to protect people, property and the environment from harm caused by hazardous materials, in particular dangerous goods and combustible liquids.

To achieve this, the Act creates broad safety obligations for all people involved with the storage, handling and manufacture of hazardous materials, together with specific obligations for occupiers and employees at locations where hazardous materials are stored or handled.

The requirements of the DGSM legislation increase as the quantity of dangerous goods stored at any premises exceeds specified amounts. Premises are classified into one of four categories as the quantity of dangerous goods increases, namely:

small quantities	→	storage workplaces other than DGLs or MHFs
medium quantities	→	dangerous goods locations (DGLs)
large quantities	→	large dangerous goods locations (Large DGLs)
very large quantities	→	major hazard facilities (MHFs) - not generally applicable in agricultural chemical situations

Most farms or persons who are contracted to spray will generally only have small quantities of agricultural chemicals on hand and will come under the provisions for minor workplaces that store or handle dangerous good, in smaller quantities. Rural workplaces are exempt from being classified as a Dangerous Goods Location or Large Dangerous Goods location if they are five or more hectares in area and are engaging in rural activities.

Some of the control measures required for rural places and other workplaces will include the following:

- obtaining and providing access to material safety data sheets for each dangerous goods and combustible liquid (e.g. diesel)
- training and supervision about safe storage and handling
- controlling ignition sources near flammable and combustible materials
- controlling and cleaning up of spills
- safe disposal of storage and handling equipment
- securing unauthorised persons from access to dangerous goods.

The following links may be of use:

An overview of dangerous goods and hazardous substances requirements at: http://www.dir.qld.gov.au/workplace/subjects

The easy to read guide on the Dangerous Goods Safety Management Act 2001 at: http://www.emergency.qld.gov.au/chem/publications/pdf/EASY_READ_GUIDE_MINE_ FINAL_SA.pdf

Safe Storage Guidelines at: http://www.emergency.qld.gov.au/chem/ publications /pdf/safe_storage_guidelines.pdf

The Workplace Health and Safety Division of the Department of Industrial Relations administers the DGSM Act and Regulation.

Health Act 1937 and Regulations

This Act brings with it the Health (Drugs and Poisons) Regulation 1996, which regulates farm chemical management.

The Health (Drugs and Poisons) Regulation X996 through the adoption of the Standard for the Uniform Scheduling of Drugs and Poisons (SUSDP) covers:

- the scheduling of individual products (poison schedules 7, 6, 5, 4)
- packaging and labelling.
- The Regulation also contains provisions that:
- restrict public access to Schedule 7 products and require their recording when sold by retail

- prohibit the use of food containers to store scheduled poisons.

In addition, there are requirements relating to the storage of Schedule 5 and 6 poisons with respect to ensuring that they are kept well away from the reach of children, as directed by the signal heading on the agricultural chemical product label.

The Department of Health administers the Health Act 1937 and Regulations.

Environmental Protection Act 1994 (EP Act) and Regulation

The Environmental Protection Act 1994 (EP Act) with its subordinate regulations and policies is primarily concerned with preventing the off-farm impact of agricultural chemicals.

The EP Act requires every person to display a general environmental duty, and not to carry out any activity that causes, or is likely to cause, environmental harm unless all reasonable and practicable measures are taken to prevent or minimise the harm. An activity (e.g. release of agricultural chemicals) that causes serious or material environmental harm or creates an environmental nuisance is unlawful unless authorised by the Act.

The Environmental Protection (Water) Policy 1997 prohibits the release of any agricultural chemical to stormwater drains or waterways, although use of an agricultural chemical in accordance with label instructions (e. g. for weed control) is considered a defence against any prosecution.

The Environmental Protection (Waste Management) Regulation 2000 prohibits the dumping of waste agricultural chemicals or containers, other than at approved waste management facilities. Waste agricultural chemicals are 'regulated wastes' and their handling and disposal is subject to specific approvals and conditions relating to their transport, handling and disposal.

The Environmental Protection Regulation 1998 prohibits environmental nuisance caused by the emission of fumes and odour. Nuisance complaints must be investigated and responded to by the administering authority.

The Environmental Protection Agency administers the EP Act and Regulations.

HERBICIDE MODES OF ACTION GROUPS

Group A: High Resistance Risk

This group contains all the 'fops' and 'dims' (so called after the end syllables in their common chemical names, e.g. haloxyfop and sethoxydim), which are selectively active on grasses in broadleaved crops. They work by reducing the ability of the plant to manufacture fat. Sugars and fats are interchanged in plants to become either energy sources for immediate growth or reserves for the future.

Group B: High Resistance Risk

These are active at very low rates (grams) per hectare. The chemical families are the sulfonylureas, the imidazolinones and the sulfonanilides. These herbicides inhibit the plant's ability to manufacture certain branched amino acids, which effectively block cell division, as these are the building blocks needed for growth.

Group C: Moderate Resistance Risk

A large collection of chemical families induces the triazines, substituted uteri, nitrites and several smaller groupings. They are all capable of interfering with photosynthesis. Without photosynthesis, and the sugars it produces, green plants are starved of energy.

Group D: Moderate Resistance Risk

Chemicals in this group had their origin in the dye industry. They are called the dinitroanalines and include trifluralin, pendimethalin and chlorthal-dimethyl. They act by preventing an early part of the cell division process. These soil-applied herbicides do not travel far in the plant and they affect root development. They stop the growth of lateral roots. Without an adequate root system, the plant is unable to sustain upward growth and it dies.

Group E: Moderate Resistance Risk

This group is made up of the thiocarbamates such as EPTC (Eptam) and tri-allate (Avadex). They are all soil-active, highly volatile chemicals that need immediate incorporation after application to avoid loss by evaporation. They are active on a range of weeds. Each member of the group has selectivity in different crops. Their killing action stops cell division, but they act at a slightly later stage than Group D.

Group F: Moderate Resistance Risk

This group includes three chemical families and selective herbicides such as diflufenican (Brodal) and norflurazon (Solicam). Their action prevents the formation of the yellow or red pigments responsible for trapping certain light wavelengths used in photosynthesis, therefore hindering the process.

Group G: Moderate Resistance Risk

This group includes the herbicides aciflurofen (Blazer), oxyflurofen (Goal) and oxadiazon (Ronstar). They act as a contaminating film (layer) on the soil surface. Plants growing through the film contact the herbicide and die. The biochemical action prevents chlorophyll manufacture without which no photosynthesis can occur.

Group H

Only one chemical appears in this division-thiobencarb (Saturn), a selective rice herbicide. It interferes with the polymerisation or joining together of amino acids that are the basic building blocks of protein.

Group I: Low Resistance Risk

This large grouping contains all the hormone growth-regulators such as 2, 4-D, MCPA, dicamba, triclopyr and other phenoxy types. They all act at a number of sites in the plant to influence the growth and development of cells by mimicking naturally occurring growth hormones in a disruptive manner. New growth appears in grotesque shapes and is often accelerated to a level that cannot be sustained by the plants internal transport system. A slow death occurs as systems gradually cease to function. Members of this group produce characteristic shapes in the plants they affect.

Group J: Low Resistance Risk

One of the members of this group, TCA, acts by accumulating in target plants and reacting with protein when concentrations reach a high enough level.

Group K: Low Resistance Risk

Described as having 'diverse sites of action', this group contains a miscellaneous number of herbicides such as the chloroacetanilides, e.g. metolachlor (Dual); the organoarsenic compounds, (e.g. DSMA and MSMA); carbamates, e.g. asulam (Asulox) and nitrites such as diclobenil (Casoron). They act at the cell level by disrupting the energy release process. Cell division is also inhibited.

Group L: Moderate Resistance Risk

The Bipyridyls (paraquat and diquat) make up this group. These molecules occur as two equal and opposite electrically charged halves known as ions. It is the positively charged half (the cation) that disrupts photosynthesis. As a result of their involvement, hydrogen peroxide is produced inside the chlorophyll-bearing cells. This powerful oxidising agent reacts with the nearest organic material available. This includes the membranes in the cell walls. As a result, these break down, the cell structure weakens and is unable to retain its contents. Moisture is lost and the dead tissue desiccates.

Group M: Low Resistance Risk

This group is made up of glyphosate. It is the most widely used herbicide in history. Chemically it is a phosphonate derivative of glycine. Glyphosate works by interference with the enzyme system that is responsible for the manufacture of some essential amino acids. The end result is no protein being made, no growth sustained and eventual death.

Group N

Represented by the increasingly important herbicide glufosinate-ammonium (Basta). The active molecule inhibits a number of important reactions that are vital to photosynthesis. The consequent accumulation of ammonium ions in the cells inhibits this process.

HOW TO MANAGE HERBICIDES WITH DIFFERENT MODES OF ACTION - SOME EXAMPLES

The classification of herbicides is based on their mode of action. It can be used as a guide to avoid repeated use of similar herbicide products. This lessens the chances of resistance problems arising. Herbicides have properties determined by their chemical nature and the way they are formulated. A selection of five of the groups is given below. **Note: in every case, the instructions for getting best results are to be found on the label.**

Group A: Example-sethoxydim formulated as Sertin 186 EC

This herbicide is used for selectively killing grasses in a large number of broadleaved crops. It is recommended that sethoxydim be applied in water volumes up to 400 L/ha and that a crop oil or approved non-ionic surfactant be added to the spray mixture, which indicates: a) good coverage is needed, and b) a little assistance in penetrating the target is required. Speed of uptake is relatively rapid, being complete in about two hours. Once in its target, sethoxydim moves well both through the phloem and the xylem and goes to work where cell division is most active.

The result is that growth stops quickly. The younger leaves turn a pale green and yellow and older growth turns a characteristic red and purple colour. Residues in the soil usually degrade within a day or so depending on the temperature, the moisture available and the pH. Fastest breakdown is at 20°C + in slightly moist acidic soil. Residues may persist longer under alkaline conditions, but in neither case can give any useful weed control effect. Best results are obtained with non-stressed weeds and the addition of a surfactant.

Group C: Example-diuron formulated as Diuron 500 SC

Diuron is a broad-spectrum herbicide that can be applied both pre- and post-emergent. It can be used selectively by applying different rates and using directed sprays in row crops. Diuron is more effective when it enters weed targets through the roots, as its movement in the plant is in the xylem only. If applied post-emergent, a wetting agent is needed to assist leaf penetration and there is little movement from where it lands on the leaf. This means good coverage must be obtained. The use of a wetter will reduce selectivity. High levels of organic matter can tie-up diuron and make it unavailable. Allowance must be made for this in non-crop situations where dead plant material has accumulated.

Diuron acts slowly when applied to the soil but its symptoms are clear when it starts working. Affected plants turn a pale yellow starting at the leaf tips and moving towards the centre of the plant. As this occurs the leading edges of the leaves die at the margins. This gradually spreads until the plant collapses. Diuron deposits are sensitive to sunlight exposure and, where weed root uptake is the objective, a little rain after application assists in obtaining a good result. Note that diuron has low water solubility and this formulation (a suspension concentrate) needs constant agitation in the spray tank to prevent it dropping out of suspension and to ensure an even dose is applied. Suspended diuron particles in the formulation also cause accelerated nozzle wear by abrasion.

Group I: Example-2,4-D formulated as Amicide 500

This herbicide contains 500 g/L of 2,4-dichlorophenoxyacetic acid (2,4-D), present as the dimethylamine salt. It is a water-based true solution. Group I includes a number of products that have the ability to disrupt a plant's normal growth. Some are better suited to specific weed families, some work best on woody weeds, some are particularly good at killing leguminous species and so on. It is not unusual for members of this group to be used in mixtures where a number of different weed groups appear in a community. 2,4-D amine salt is a herbicide that exhibits selectivity in some monocotyledon crops such as winter cereals. But its safety in use depends on getting both the dose applied and the timing relative to crop development right.

Some broadleaved crops are extremely sensitive to small amounts of 2,4-D and application must always ensure there is no possibility of drift occurring if there are any such crops in the area (Tomatoes, papaws and brassica vegetables are among the most sensitive). The more volatile ester formulations are banned or their use restricted in certain designated mixed cropping areas as there is a real danger that sufficient of the ester deposits will evaporate on a hot day to cause damage to sensitive crops if transferred in the wind.

The water-based amine salt is not as influenced by temperature as the ester but still needs careful management. 2,4-D is slow to penetrate its target and it takes about four hours to be sure a lethal dose has been absorbed, i.e. no rain is wanted within this period after spraying. Once in the target, it moves well in all directions and disorganises the plant's normal growth processes. 2,4-D in the soil is degraded by micro-organisms but, depending on the dose, residues can persist for four to five weeks. 2,4-D is only weakly adsorbed to the clay fraction in the soil, which means it may leach through the soil profile to ground water and calls for care especially on light soils where physical downward movement after heavy rain is possible.

Group L: Example - paraquat formulated as Gramoxone 250

A water-based formulation containing 250 g/L of paraquat, as the dichloride salt. This product shares with diquat the distinction of being taken up by its target more quickly than any other herbicide grouping. Within five to ten minutes of hitting its target, the herbicide is completely weatherproof and is not vulnerable to rain-wash. This happens because the active part of the ionised molecule is a strong positively charged cation that is adsorbed by the negatively charged (earthed) plant.

A similar effect occurs when paraquat touches the soil that results in it being strongly bound to the clay fraction in the soil and becoming biologically unavailable to all but a few bacteria. This means there is instantly a non-residual situation, which in turn means crops can be seeded immediately following a spray although the normal recommendation is to delay two or three days. As paraquat is inactivated by the soil, any water used for making up spray solutions must be clean and free of any soil contamination.

Paraquat starts to work where it hits its target and does not move to any extent after it is locked into the process of photosynthesis. This makes the use of surfactants essential to ensure a good cover of the sprayed weeds. (It is important that a non-ionic or neutrally charged surfactant be used. The use of anionic or negatively charged surfactants must be avoided to prevent reaction and subsequent loss of activity within the spray tank).

If photosynthesis is not taking place (at night) or is slowed down due to overcast conditions, some movement into the plant can occur. In a situation where good coverage is hard to achieve because of dense-masking vegetation, spraying with a high volume towards the in full sunlight. Paraquat is non-selective and non-residual and under sunny conditions can complete its desiccating action in four or five days. Paraquat tends to have more activity on grasses than broadleaved weeds whereas with diquat, the opposite is true. As a result, mixtures of the two are formulated to broaden the weed control spectrum (e.g. Spray Seed). Diquat is more light-sensitive than paraquat and breaks down completely within the few days it takes to desiccate its target. This makes it the chemical of choice for pre-harvest desiccation sprays.

Group M: Example - glyphosate formulated as Roundup

A water-based formulation containing 36o g/L of glyphosate, present as the isopropylamine salt. There are several different salts of glyphosate on the market but their effects are all similar. Glyphosate is, like paraquat, non-selective at recommended rates and inactivated on contact with the soil. The use of clean water and non-ionic surfactants apply here also. Glyphosate is more sensitive to the pH of the water with which it is mixed and appears to be more stable and to work more consistently when the water is acid. This aspect is particularly important when mixtures with other chemicals are recommended and acidifying buffer solutions are available to assist in reducing mixture antagonism. Glyphosate can take up to six hours to penetrate its target although there has been much progress with newer formulations to reduce this period. Rain within this period may mean a poor result.

Once in the plant, glyphosate translocates quickly. It is useful to time the spray to take advantage of any downward movement of photosynthesis products from the leaves to make sure the glyphosate travels right down to the roots and storage organs of the weed. With hard-to-kill weeds, like nut grass, the timing can be critical. The time taken for glyphosate to complete its task varies with the species and the degree of stress in the target. But, under good conditions, competition from the sprayed weeds stops in a matter of days and total death occurs in a couple of weeks.

AVCARE GROUPS (MODES OF ACTION) FOR MANAGING HERBICIDE RESISTANCE

High risk

Group A

Fops	Correct, Fusilade, Fusion, Shogun, Targa, Topik Tristar Advance, Spear, Wildcat, Puma Progress, Decision, diclofop, haloxyfop
Dims	Achieve, Falcon, Fusion, Select, Sertin Plus, Sertin 186, Aramo

Group B

Sulfonylureas	Cut.OutTM (also contains glyphosate Group M), Harmony M, Hussar, Londax, Monza, Oust, Titus, metsulfuron, chlorsulfuron, triasulfuron, Envoke, Atlantis, Bocnty (also contains diflufenican Group F)
Imidazolinones	Arsenal, Flame, Lightning, Midas (also contains MCPA Group I), OnDuty, Raptor, Spinnaker
Sulfonamides	Broadstrike, Eclipse

Moderate risk

Group C

Triazines	Agtryne MA (also contains MCPA Group I), atrazine, simazine, terbutryn, cyanazine
Triazinones	Velpar, metribuzin
Ureas	Graslan, Tribunil, Tupersan, diuron, linuron
Nitriles	Jaguar (also contains diflufenican Group F), Barracuda (also contains diflufenican Group F), bromoxynil, ioxynil

Benzothiadiazoles	BasagranM60 (also contains MCPA Group I), Basagran
Acetamides	Propanil
Uracils	Hyvar, Krovar, Sinbar
Pyridazinones	Pyramin
Group D	
Dinitroanilines	Surflan, Yield, trifluralin, pendimethalin
Benzoic acids	Dacthal, chlorthal-dimethyl
Group E	
Thiocarbamates	AvadexXrta Partner, Eptam, Saturn, Tillam Vernam, molinate
Carbamates	Chlorpropham
Organophosphorous	
Group F	
Nicotinanilides	Brodal, Jaguar (also contains bromoxynil Group C), Barracuda (also contains bromoxynil Group C), Tigrex (also contains MCPA Group I), Giant((also contains MCPA Group I), Bounty (also contains ethametsulfuron Group B)
Triazoles	amitrole
Pyridazinones	norflurazon
Isoxazolidinones	Command, Magister, Viper (also contains benzofenap Group F)
Pyrazoles	Taipan, Viper (also contains clomazone Group F)
Picolinamides	Paragon (also contains MCPA Group I), Sniper
Isoxaflutole	Balance
Group G	
Diphenylethers	Blazer, oxyfluorfen
Oxadiazoles, Triazolinones, Pyrimidindione	Ronsatar Affinity AffinityPlus (also contains MCPA Group I), Hammer LogranB Power (also contains triasulfuron Group B)

Low Risk

Group I	
Phenoxys	2,4-D, 2,4-DB, MCPA, Barrel (also contains bromoxynil Group C and dicamba Group I), Buctril MA (also contains bromoxynil Group C), Midas (also contains imazapic/imazapyr Group B), Paragon (also

	contains picolinofen Group F), Tigrex (also contains diflufen can Group F), Giant (also contains diflufenican Group F) BasagranM60 (also contains bentazone Group C)
Benzoic acids	dicamba
Pyridines	Grazon DS, Lontrel, Tordon242, Tordon 75-D, triclopyr, fluroxypur
Group J	
Alkanoic acides	2, 2-DPA, flupropanate
Group K	
Amides	Devrinol, Dual Gold, metolachlor
Carbamates	Betanal, asulam
Amino propionates	Metaven L
Benzfurans	ethofumesate
Nitriles	Casoron
Dicarboxylic acid	endothal
Group L	
Bipyridyls	Spray Seed, Regione, Paraquat
Group M	
Glycines	Touchdown Hi-Tech, glyphosate, Roundup Ready Herbicide, RoundupPowerMax
Group N	
Phosphinic acids	Basta

Section - D

MODES OF ACTION OF INSECTICIDE GROUPS

Insecticides are classified by numbered groups, with letters being used to identify sub-groups.

Biologicals or bio-control agents

Public concern with synthetic agricultural chemicals has encouraged the introduction of environmentally benign agents isolated from nature and used for pest control either alone or alongside synthetic compounds in integrated pest management (IPM) programs. The number of marketable natural agents available is growing steadily as products are sought that can be used as alternatives to the organophosphates and carbamates. Examples may be found in Groups 5, 6, and 11 (see Appendix E).

One of the most established is Bacillus thuringiensis (B.t.). This bacterium has the ability to form toxic protein crystals that dissolve and mobilise in the alkaline gut of some pest moth larvae including Helicoverpa spp. The result is an alteration of the permeability of the gut wall that allows the spread of the toxin resulting in eventual death. There are several strains of B. t. which appear to have their activity linked to different target larvae. Another useful bacterium Saccharopolyspora spinosa with the approved name of spinosad produces breakdown chemicals that attack the nervous system. Viruses are also becoming available and an example is a baculovirus with the approved name Helicoverpa zea nucleopolyhederovirus, usually shortened to HzSNPV or HzNPV. Yet a Bother example is abamectin, which is the result of fermenting a naturally occurring soilactinornycete Streptomyces avermitilis. This bio-agent inhibits the transmission of messages between the nervous system and the muscles.

Group 1A: Carbamates

The carbamates are toxic insecticides that act quickly but often degrade quickly.

Carbamates, like the organophosphates, work by inhibiting the release of cholinesterase at nerve synapses, therefore allowing the build-up of acetylcholine between successive nerve cells. As a result, instead of 'pulsed' brain messages reaching muscles to be activated, a continuous stimulation passes down the line, eventually causing muscular spasm and incoordination. As the action in insects and humans is similar, these compounds qualify for the label warning: 'An anti-cholinesterase compound'.

Group 1B: Organophosphates

The organophosphates (OPs) have their origins in Germany in 1938 and have served agriculture well ever since. The mode of action is similar to the carbamates and the allocation of the two groups to two segments is probably due to differences in chemistry within the target while moving towards the site of action in the nervous system. Some OPs need biochemical activation to enable them to inhibit the release of cholinesterase.

Groups 2A, 2B and 2C: GABA-gated chloride channel antagonists

Compounds listed in this category include dicofol (Kelthane) and tetradifon (Tedion), two miticides of long standing, and endosulfan. Also in the category is fipronil (Regent) that acts in a similar electron-blocking manner. These chemicals have a mode of action similar to that of the synthetic pyrethroids. They attack the nervous system but by a different mechanism to the organophosphates and carbamates. Their site of action is at the end of the transmission pathway along the nerve network where contact is made with the muscle to be activated.

Group 3A: Synthetic pyrethroids

This large and important group was launched in the mid 1970s after nearly 30 years of research to achieve light stability, which was missing in the first natural pyrethroids. The molecules are synthetic copies imitating the naturally occurring pyrethrins extracted from the flower Chrysanthemum cinerariaefolium. The synthetic pyrethroids act on the nervous system by accumulating in the myelin sheath, which is a protective layer of fatty tissue surrounding the elongated part of the nerve cell known as the axon. Brain-to-muscle messages pass along the axon as a ripple of movement of sodium ions in both directions, in and out of the membrane surrounding the axon. The presence of the pyrethroids interrupts this sequence of movements so the message being transmitted is blocked or disrupted and becomes scrambled. The end result is muscula- incoordination. Responsible management of the group is needed to ensure synthetic pyrethroids are not over applied with consequent loss of usefulness due to resistance.

Group 4A: Chloro-nicotnyls

There are four current members of this group, acetamiprid (Intruder, Mospilan) imidacloprid (Confidor), thiacloprid (Calypso) and thiamethoxam (Actara, Meridian). These insecticides work by irreversibly blocking the post-synaptic acetylcholine receptors. They act as both a contact and a stomach poison and have good systemic action in plants and good activity against aphids. Chemically, they slow up the passage of nerve impulses as blocked receptors and the message being transmitted comes to a halt.

Groups 7A, 7B and 7C: Juvenile hormone mimics

The insecticides in this group all have the ability to disrupt processes in metamorphosis so that the targets are unable to mature properly and die before being able to reproduce.

Groups 8A and 8B: Fumigants

The first of these is methyl bromide that has been used as a soil and general fumigant for nearly 70 years. It is now being slowly phased out as it contributes towards ozone-layer destruction. The second segment includes all the aluminium phosphide products that generate phosphine gas on exposure to moisture at low levels. The products are placed in stored cereals and the moisture in the grain is adequate to generate phosphine gas in quantities adequate to kill all grain pests present.

Groups 9A and 9B: Unknown or non-specific modes of action

There is one registered agricultural chemical within these groups-pyrnetrozine (Chess). It has good activity against sucking insects and mites. Its selective action makes it especially useful in IPM programs.

Group 9A: Miticides (early stages)

Two products in this group, clofentezine (Apollo) and hexythiazox (Calibre), have good action on eggs and young stages of pest mites, but little effect on adults. They interfere with embryonic development. Conversely, pyridaben (Sanmite) and tebufenpyrad (Pyranica) are active against all stages of mite development and, as both contact and stomach poisons, have rapid knockdown action.

Group 12A and 12B

Members of these two groups act on a wide range of adult mite species as both contact and stomach poisons. They include the organo-tin compounds such as fenbutatin oxide (Torque), as well as the substituted urea compound diafenthiuron (Pegasus). Both products have a similar mode of action. They are relatively safe to predatory mites and are sometimes used as bio-control agents. They are thought to act by inhibiting the processes involved in the pests' storage and release of energy.

Group 13A

The sole member of this group is chlorfenapyr (Secure). An insecticide and acaricide with stomach and some contact action, it has useful application in some integrated pest management programs in fruit orchards.

Group 14A: Sulfite ester miticides

Propargite is within this group and is sold under two trade names, Comite (for cotton) and Omite (for orchards), depending on its market use. This miticide is yet another single classification. The detailed mode of action is not known.

Group 15A: Chitin inhibitors

There are currently three members (diflubenzuron, hexaflumuron and triflumuron) with this mode of action. They prevent the growth of the new outer skin of an insect after a moult, making it vulnerable to both desiccation and microbial attack.

Group 16A: Ecdyson agonists

The two current members of this group are methoxyfenozide (Prodigy) and tebufenozide (Mimic). They lethally accelerate the moulting process of the target insects.

Group 17A

The sole member of this group is buprofezin (Applaud). Its probable mode of action is as a chitin synthesis and prostaglandin inhibitor. It has a hormone-disturbing effect, leading to suppression of ecdysis (moulting).

Group 18A: Insect growth regulator

The veterinary chemical cyromazine (Vetrazine) affects both the moulting and pupation processes preventing normal progress to maturity.

Group 19A: Multi-function insecticide, miticide and tickicide

The sole member of this group is amitraz that is widely used as an ovicide of Helicoverpa spp. eggs as well as a control of cattle tick.

Group 20A

There is currently only one member of this group-hydramethylnon (Amdro). It is non-systemic with stomach action and finds use as a bait treatment against urban pests, particularly ants.

Group 21A

Rotenone, extracted from derris root, is currently the sole entry in this grouping.

Group 22A

A group also currently containing a sale member-indoxacarb (Avatar, Steward). It has a unique mode of action. Enzymes in the gut of the insect convert the chemical to the active metabolite that then paralyses its nervous system. It is finding use in insecticide resistance management and IPM programs.

Unspecified

This miscellaneous grouping contains:

- **dazomet:** A multipurpose soil fumigant including an insecticidal action. It breaks down to methyl isothiocyanate in a similar way to metham-sodium. Both are therefore potentially highly toxic to most forms of life, 'including nematodes. The poisoning effect results from interference in the nucleus energy supply enzyme system.

- **metaldehyde:** The most commonly used snail killer. It acts by increasing production of slime so that it eventually desiccates the animal by loss of water.

Some well-known older botanical insecticides

- **nicotine:** A naturally occurring alkaloid extracted from tobacco that acts by mimicking acetylcholine, but is not susceptible to cholinesterase. It kills therefore in a similar way to the organophosphates and carbamates.
- **derris root:** An extract from Derris alliptica, which acts by inhibiting respiration, which results in lower oxygen consumption, lethargy and paralysis.
- **pyrethrum:** A complex mixture of six major compounds extracted from one of the chrysanthemum family of flowers--Chrysanthemum cinerariaefolium.

AVCARE INSECTICIDE RESISTANCE ACTION COMMITTEE (AIRAC) MODES OF ACTION CLASSIFICATION FOR INSECTICIDES

Group	Primary target site	Chemical subgroups
1A	Acetyl choline esterase inhibitors	carbamates*
1B		organophosphates*
2A	GABA-gated chloride channel aitagonists	cyclodienes
2B		polychlorocycloalthanes
2C		fiproles
3A	Sodium channel modulators	pyrethroids and pyrethrins
4A	Acetyl choline receptor agonists/antagonists	chloronicotinyls
4B		nicotine
4C		cartap, bensultap
5A	Acetyl choline receptor modulators	spinosyns
6A	Chloride channel activators	avermectin, emamectin benzoate
6B		milbemycin
7A	Juvenile hormone mimics	methoprene, hydroprene
7B		fenoxycarb
7C		pyriproxifen
8A	Compounds of unknown or non-specific mode of action (fumigants)	methyl bromide
8B		phosphine generating compounds

Group	Primary target site	Chemical subgroups
9A	Compounds of unknown or non-specific mode of action (selective feeding blockers)	pymetrozine
9B		cryolite
10A	Compounds of unknown or non-specific mode of action (mite growth inhibitors)	clofentezine, hexythiazox
11A	Microbial disrupters of insect midgut membranes (includes Transgenic B.t. crops)	B.t. tenebrionis
11B		B.t. israelensis
11C		B.t. kurstaki, B.t. aizawati*
11D		B.t. sphaericus
11E		B.t. tolworthi
12A	Inhibition of oxidative phosphorylation, disrupters of ATP formation	organotin miticides
12B		diafenthiuron
13A	Uncoupler of oxidative piosphorylation via disruption of H proton gradient	chlorfenapyr
14A	Inhibition of magnesium stimulated ATPase	propargite
15A	Chitin biosynthesis inhibitors	acyl ureas
16A	Ecdysone agonists	tebufenozide and related
17A	Homopteran chitin biosynthesis inhibitors	buprofezin
18A	Unknown dipteran specific mode of action	cyromazine
19A	Octopaminergic agonist	amitraz
20A	Site II electron transport inhibitors	hydramethylnon
21A	Site I electron transport inhibitors	rotenone, METI acaricides
22A	Voltage dependent sodium channel blocker	indoxacarb

* all members of this class may not be cross resistant

Avcare Insecticide Resistance Action Committee (AIRAC) mode of action classification for insecticides: active constituent list

Table 2 (by active constituent)

Active constituent	Group	Active constituent	Group
abamectin	6A	bifenthrin	3A
acephate	1B	bioallethrin	3A
aldicarb	1A	bioresmethrin	3A
allethrin	3A	buprofezin	17A
alpha-cypermethrin	3A	cadusafos	1B
aluminium phosphide	8B	carbaryl	1A
amitraz	19A	carbofuran	1A
avermectin	6A	carbosulfan	1A
azamethiphos	1B	chlorfenvinphos	1B
azinphos methyl	1B	chlorfenapyr	13A
azinphos-ethyl	1B	chlorpyrifos	1B
Bacillus thuringiensis aizawai	11C	chlorpyrifox-methyl	1B
Bacillus thuringiensis israelensis	11B	clofentezine	10A
Bacillus thuringiensis kurstaki	11C	cryolite	9B
Bacillus thuringiensis sphaericus	11D	cyfluthrin	3A
Bacillus thuringiensis tenebrionis	11A	cypermethrin	3A
Bacillus thuringiensis tolworthi	11E	cypermethrin (zeta)	3A
bendiocarb	1A	cyromazine	18A
betacyfluthrin	3A	deltamethrin	3A
diafenthiuron	12B	lamda-cyhalothrin	3A
diazinon	1B	magnesium phosphide	8B
dichlorvos	1B	maldison (malathion)	1B
dicofol	2B	mancozeb	1A
diflubenzuron	15A	methamidophos	1B
dimethoate	1B	methidathion	1B
disulfoton	1B	methomyl	1A
emamectin benzoate	6A	methoprene	7A
endosulfan	2A	methyl bromide	8A

Active constituent	Group	Active constituent	Group
ethion	1B	mevinphos	1B
esfenvalerate	3A	milbemycin	6B
fenbutatin oxide	12A	monocrotophos	1B
fenitrothion	1B	oxamyl	1A
fenoxycarb	7B	omethoate	1B
fenpyroximate	21A	parathion	1B
fenthion	1B	parathion-methyl	1B
fipronil	2C	peremethrin	3A
flutriafol	3A	phorate	1B
fluvalinate	3A	phosmet	1B
furathiocarb	1A	phosphine	8B
hexaflumuron	15A	pirimicarb	1A
hexythiazox	10A	pirimiphos-methyl	1B
hydramethylnon	20A	profenofos	1B
hydroprene	7A	propargite	14A
imidacloprid	4A	propoxur	1A
indoxacarb	22A	prothiofos	1B
isofenphos	1B	pymetrozine	9A
pyrethrins	3A		
pyridaben	21A		
pyriproxyfen	7C		
rotenone	21A		
spinosad	5A		
sulprofos	1B		
tau fluvalinate	3A		
tebufenozide	16A		
tebufenpyrad	21A		
temephos	1B		
tetradifon	2B		
tetramethrin	3A		
thiodicarb	1A		
triflumuron	15A		
vamidothion	1B		

FUNGICIDAL MODES OF ACTION GROUPS

Group A: Benzimidazoles

Active on powdery mildews, rusts, grey mould and various leaf spots. It is believed members of this group inhibit one of the early stages of cell division. These fungicides are systemic and referred to as the MBCs-methyl benzimidazole carbamates. The original and best-known member of this group is benomyl (Benlate). It has been replaced in recent times by carbendazim (Bavistin).

Group B: Dicarboximides

Active on brown rot, grey mould, sclerotinia and apple scab. These are not truly systemic although they can have limited eradicant properties. They are thought to interfere with membrane and cell wall function in some way. The two members of this group are iprodione (Rovral) and procymidone (Sumisclex).

Group C: Demethylation inhibitors (DMI)

These are a large group of 19 fungicides subdivided into four chemical families of which the triazoles are the largest.

Some examples of members of each of the four sub-groups are:

- Imidazole - prochloraz (Octave; Sportak)
- Piperazine - triforine (Saprol)
- Pyrimidine - fenarimol (Rubigan)
- Triazole - propiconazole (Tilt)

They are collectively referred to as the DMI group, which stands for demethylation inhibitors. In practical terms, this means activity on powdery mildews, rusts, brown rot and leaf spots within the group and this is achieved because the structure and function of some of the fungal membranes is disorganised.

Group D: Phenylamides

Active against downy mildew, damping off, late blight and root rots. This group contains some very efficient systemic fungicides, such as mctalaxyl (Ridomil), which are frequently marketed as a mixture with a protectant, such as mancozeb. The mode of action is unclear but it probably is a single site activity that may interfere with RNA (ribonucleic acid) synthesis.

Group E: Morpholine (Spiroketalamine)

Active against powdery mildew. This group has only one current member- spiroxamine (Prosper). It acts in a similar way to Group C, but at different points in the biochemical sequence.

Group G: Oxathiin

Active on rusts and smuts. Two members of this group are carboxin (Vitavax), a seed dressing and oxycarboxin (Plantvax), an eradicant rust spray. Their mode of action is not clear.

Group H: Hydroxy-pyrimidines

Active on powdery mildew. This group contains the fungicide bupirimate (Nimrod), which inhibits the enzyme system involved in nucleic acid synthesis.

Group I: Anilinopyrimidines

The two members of this group cyprodinil (Chorus) and pyrimethanil (Scala) are recent introductions with only limited registrations. Their mode of action differs from other groups with a similar fungicidal spectrum that makes them useful in resistance management.

Group J: Hydroxyanilide

Another new group with only a single member-fenhexamid (Teldor).

Group K: Strobilurins

Currently contains four chemicals - azoxystrobin (Armistar), kresoxim-methyl (Stroby), pyraclostrobin (Cabrio) and trifloxystrobin

(Flint). They are the result of a decade of research into modifying an isolate from a natural fungus to improve its activity and stability. They are the first fungicides to show activity on both powdery and downy mildew as well as attacking several other fungal families.

Group Y: Mufti-site activity

This wide-ranging group includes some of the most frequently used fungicides, such as the copper-based salts and dithiocarbarnates (e.g. Dithane, Antracol). The group contains sulphur, chlorothalonil (Bravo) and some others. The interaction between some of these surface acting fungicides and other formulations, particularly those containing oils or solvents capable of penetrating cuticle wax, has to be managed with care to avoid leaf damage. Some of the chemicals within this group are used in mixtures with systemic fungicides, as it seems that the dual effect produced not only improves effectiveness but also lessens the chance of resistance selection.

Group X: Unspecified

This classification is an indication of the limited knowledge about how fungicides work. It contains eight different chemical families including dimethomorph (Acrobat), oxythioquinox (Morestan) and tolclofos-methyl (Rizolex).

AVCARE FUNGICIDAL MODES OF ACTION AND ACTIVITY GROUPS

Group	Activity group	Chemical grouping	Active constituent	Trade name
A	Benzimidazole	Benzimidazole	benomyl	Benlate
			carbendazim	various e.g. Bavistin
			thiabendazole	Tecto
			thiophanate-methyl	
B	Dicarboximide	Dicarboximide	iprodione	Rovral
			procymidone	Sumisclex, Fortress
C	DMI	Imidazole	imazalil	various e.g. Fungaflor, Magnate
			prochloraz	various e.g. Octave
		Piperazine	triforine	Saprol
		Pyrimidine	fenarimol	Rubigan
		Triazole	bitertanol	Baycor
			cyproconazole	Alto
			cyproconazole (+ iodocarb)	Garrison
			diclobutrazole	Vigil
			difenoconazole	Score, Bogard
			epoxiconazole	Opus
			fluquinconazole	Jockey
			flusilazole	Cane Strike, Nustar
			flutriafol	various e.g. Armour
			hexaconazole	Anvil
			myclobutanil	Systhane, Mycloss
			paclobutrazol	Cultar
			penconazole	Topas
			propiconazole	various e.g. Tilt

Group	Activity group	Chemical grouping	Active constituent	Trade name
D	Phenylamide	Acylamine	benalaxyl (+ mancozeb)	GalbenM
			furalaxyl	Fongarid
			metalaxyl (+ mancozeb)	RidomilMZ
			metalaxyl - m	RidomilGOLD
			metalaxyl-m (+ copper oxychloride)	Ridomil GoldPlus
			metalaxyl-m (+ mancozeb)	RidomilGoldMZ
		Oxazolidinone	oxadixyl (+ mancozeb)	Recoil
			oxadixyl (+propineb)	Fruvit
E	Morpholine	Morpholine	tridemorph	
		Spiroketalamine	spiroxamine	Propser
F	Phosphoro-thiolate	Organo-phosphorous	pyrazophos	Afugan
G	Oxathilin	Anilide	carboxin	Vitavax, Regatta
			flutolanil	Moncut
			oxycarboxin	Plantvax
			boscalid	Filan
H	Hydroxy-pyrimidine	Pyrimidinol	bupirimate	Nimrod
I	Anilinopyrimi-dine	Anilinopyrimidine	cyprodinil	Chorus
			cyprodinil (+ fludioxinil)	Switch
			pyrimethanil	Scala
			pyrimethanil (+chlorothalonil)	Walabi
J	Hydroxyanilide	Hydroxyanilide	fenhexamid	Teldor
K	Strobilurin	Strobilurin	azoxystrobin	Amistar
			azoxystrobin (+cyproconazole)	Amistar Xtra
			kresoxim-methyl	Stroby
			pyraclostrobin	Cabrio
			trifloxystrobin	Flint

Group	Activity group	Chemical grouping	Active constituent	Trade name
L	Phenylpyrroles	Phenylpyrroles	fludioxonil	Maxim
M	Phenoxy quinoline	Phenoxy quinoline	quinoxyfen	Legend
Y	Multi-site activity	Carbamate	Iodocarb	
			propamocarb	Previcur
		Phosphonate	fosetyl-al	Aliette
			phosphorous acid	various
		Inorganic Dithiocarbamate	chlorine dioxide	Vibrex Flora
			copper cuprous oxide	various
			copper hydroxide	various
			copper oxychloride	various
			copper octanoate	Tricop
			hydrogen peroxide + peroxyacetic acid	Peratec
			iodine	Ultra Dyne C
			mercury	Shirtan
			sodium metabisulphite	Uvas Quality Grapeguard
			sulphur	various e.g. Thiovit, Kumulus
			mancozeb	various e.g. Dithane
			thiram	various e.g. Thiram
			propineb	Antracol
			zineb	various e.g. Zineb
			zineb (+ copper oxychloride)	Copper Curit
			ziram	various e.g. Ziram
		Phthalimide	chlorothalonil	various e.g. Bravo
		chlorophenyl	quintozene	Terrachlor
		Quinone	dithianon	Delan
		Hydroxyquino-line	8-hydroxy quinoline sulphate	Chinosol
		Pyradinamine	fluazinam	Shirlan
		Cyclic imide	captan	Capan, Merpan

Group	Activity group	Chemical grouping	Active constituent	Trade name
X	(Unspecified)	Cinnamic acid derivative	dimethomorph	Acrobat
		Sulfamide	dichlofluanid	Euparen
			tolyfluanid	Euparen Multi
		Dinitrophenyl	dinocap	Karathane
		Organophosph-ate	tolclofos-methyl	Rizolex
		Guanidine	dodine	various e.g. Dodine
			guazatine	Panoctine
		Thiadiazole	etridiazole	Terrazole
		Quinoxaline	oxythioquinox	Morestan
			pencycuron	Monceren

Note :

Some products are mixtures of fungicides from different activity groups. These appear only once in the chart. If multiple trade names exist, the trade name entry is listed as various and a common tradename included.

GLOSSARY

Acaricide : An agricultural chemical for killing mites and ticks (Order Acarina) as distinct from insects.

Active constituent or ingredient : The biologically active part of the agricultural chemical present in a formulation. The two terms Active Constituent and Active Ingredient are often synonymous, but the former term is used on the product label. Sometimes the difference in meaning can be due to the fact that the chemical in question is produced in forms differing from the original base material (e.g. 2,4 D).

Acute oral toxicity : A measure of the level of toxicity or ability to poison by ingestion of a chemical. It is expressed in milligrams of the chemical per kilogram liveweight of the test species being used as a standard in feeding trials.

ADG Code : The Australian Dangerous Goods (ADG) Code that provides a guide for the safe transport of chemicals classified by the United Nations as being Dangerous Goods (DG) because of their potentially hazardous properties.

Adjuvant : Any substance aided to an agricultural chemical formulation to make it work better. Examples include adhesives, emulsifiers, penetrants, spreaders and wetting agents.

Adsorption : An increase in the concentration of a dissolved substance at the interface of a solid and a liquid phase due to the operation of surface forces.

Adventitious roots : Roots arising elsewhere than from the taproot, e.g. roots along a stem, or those developed from a stem or leaf cutting.

Agricultural chemical : A broad term used to cover pesticides, adjuvants, plant growth regulators, defoliants and other chemical tools used in improving agricultural production, protecting crops or controlling pests, diseases and physiological conditions of crops or plants. The term does not normally include fertilisers.

Agricultural chemical formulation : The combination of the various ingredients (active and inert) that comprise the agricultural chemical product as sold. Most are prepared in concentrated forms that require dilution in a carrier such as water before application to the crop, plant or target area.

Anionic : Negatively charged (ions).

Annuals : Used to describe plants that complete their life cycle within a single year.

Anticholinesterase compound : A chemical with the ability to inhibit the production of the enzyme cholinesterase, therefore allowing the build-up of acetylcholine between individual nerve cells and muscle cells. As a result, muscles are continuously stimulated and become uncoordinated. The insecticides in the organophosphate and carbamate groups are the best-known examples.

Antidote : Medicine taken or administered to counteract the effects of poisoning.

Aphicide : Agricultural chemical used for the control of aphids.

Arachnida, Arachnids : Class of arthropods that include spiders and mites.

Arthropods : A very large phylum of invertebrate animals that are characterised by having exoskeletons and jointed appendages (limbs). The grouping includes all insect and arachnid pests and beneficial species.

Authorised person : A person authorised to purchase and use restricted agricultural chemical products such as endosulfan.

Avcare : Avcare is the national association for crop production and animal health in Australia. It represents manufacturers, formulators and distributors of crop protection, animal health and biotechnology products.

Bactericides : Chemicals used to kill bacteria, sometimes referred to as sanitisers. They are often anionic surfactants.

Beneficial insects : Any insect that is useful or helpful to humans (e.g. honeybees). Some beneficial insects attack pest species that inflict economic damage to crops.

Biennials : Plants that take two years to complete their life cycle.

Bio-control agent : A living species with the ability to attack and kill pests. It may be a vertebrate or invertebrate animal or a fungal organism.

Biodiversity : A number of different biological species present in a nominated environment, e.g. a farm or plant nursery.

Biomass : The total weight or volume of living material present in an individual of area.

Brassica vegetables : Vegetables in the cabbage family, including broccoli, brussels sprouts, etc.

Buffering agent : In chemistry, a chemical which, when added to a solution, has the ability to resist changes to pH or hydrogen ion concentrations. Acidifying buffers are used to counter alkaline bore water supplies that can improve spray solution stability and performance.

Bulbs : A plant storage and survival organ consisting of a short stem tightly surrounded by swollen overlapping modified leaves. Usually below ground but sometimes exposed, as with onions.

Cationic : Positively charged (ions).

Chlorophyll : The green pigment in plants responsible for photosynthesis, a complex process that uses sunlight as an energy source to convert carbon dioxide and water molecules into sugars and oxygen.

Chronic toxicity : The ability of a chemical to cause injury by repeated exposure to small amounts over a period of time.

Coleoptera : An order of insects, including beetles and weevils.

Contact poison : A poison that cases injury on contact, e.g. a herbicide that acts where it hits a weed with little or no movement within the plant. An insecticide that acts without having to be ingested.

Corms : An enlarged solid underground stem, which contains dormant buds for future growth. Usually round in shape and covered by thin scaly leaves.

Curative fungicides : Fungicides that are able to move within the plant being attacked by a fungus and selectively kill or inhibit that fungus. Also called eradicant fungicides.

Cuticle : The outer covering of leaves, which is often a waxy protective layer.

Desiccation : The process of losing moisture and withering.

DG Class Labels : The diamond-shaped colour-coded symbols that indicate the Dangerous Goods (DG) class.

Dicotyledon (dicots) : A sub-class of plants characterised by having two first seed leaves or cotyledons. Most broadleaved plants fall within this category.

Diffusion : The spreading and penetration of particles by natural movement into space that may be enclosed, as in the case of respirator filter elements.

Diptera : An order of insects that includes flies, sandflies and mosquitoes.

Disinfectant : A chemical applied to kill micro-organisms, usually bacteria.

Dispersal : The process of spreading a population usually by seeds or spores.

Emulsifiable concentrate : An agricultural chemical formulation consisting of an active constituent dissolved in an organic solvent together with an emulsifier to facilitate the formation of an even milky emulsion when mixed with water.

Environmental fate : The term used to describe the breakdown pathways of agricultural chemicals after they have been applied.

Environmental load : The total amount of chemical put into the environment when an agricultural chemical is applied.

Ephemerals : Plants that have rapid generation succession with several generations following each other in a favourable season.

Eradicant fungicides : See Curative fungicides.

Exoskeleton : The hard covering that characterises arthropod bodies-an 'outside skeleton'.

Flowable formulation : Sometimes called suspension concentrates. These are formulations of finely milled solid active particles suspended in a liquid carrier (usually water).

Formulation : The make-up of the agricultural chemical product as purchased. It consists of the active constituent(s) together with a number of other components that are added to assist handling, efficacy, safety and stability.

Fumigation : The process of using toxic vapours to penetrate sealed-off produce or soil or enclosed areas to kill any pest organisms present.

Fungi

(singular fungus) : Non-chlorophyll-bearing plants e.g., rusts, mildews and moulds. Some are parasites and can cause diseases in plants and animals. Others are beneficial and provide food sources (e.g. mushrooms) or help degrade dead plant or animal matter.

Fungicidal dip : A high volume tangy: of fungicide solution used to treat plant material either pre-planting or post-harvest to control any pathogenic fungi present and give short-term protection.

Fungicide : Any substance or mixture of substances intended for preventing, destroying or controlling any fungi.

Hazchem Code : A system of numbers and letters that indicate the appropriate actions to take in the event of a chemical emergency. They are displayed on bulk transport vehicles and the entrance of storage areas when required by legislation.

Hemiptera : An order of insects that includes ones with sucking mouthparts adapted to feed on plant sap. Examples are aphids, bugs, leafhoppers and scales.

Herbaceous weeds : Weeds that have an abundance of soft green foliage and dc not develop woody tissue.

Herbicide : Any substance or mixture of substances intended for preventing, destroying or controlling any unwanted plant, including any algae or aquatic weed.

Humectant : A hygroscopic (water-attracting) chemical.

Hypha(e) : The basic structure of the body or mycelium of fungi consisting of a filament-like tube-shaped body of cytoplasm containing nuclei that may or may not be divided into segments by septa (walls).

Inert ingredient : Any ingredient in an agricultural chemical formulation that has no inherent biological activity against the target pest or disease, as opposed to the active constituent or active ingredient.

Ingestion : Eating or taking into the body via the mouth.

Inhalation : Breathing, drawing in air to fill the lungs.

Insect growth regulator : Chemicals capable of influencing the biochemical processes involved during the transition of an insect from one life cycle phase to another, e.g. the inhibition of chitin formation after a moult

Insecticide : Any substance or mixture of substances intended for preventing, destroying, repelling or controlling any insect at any stage of its development, e.g. eggs, instars, larvae, nymphs, pupae or adults. The term often includes substances used similarly against allied classes of arthropods, such as spiders, mites, ticks, centipedes, millipedes and slaters.

Instar : The word used to describe a growth stage of an insect between moults. Therefore, a third stage instar larva of a moth would be the caterpillar between the second and third moult.

Integrated pest management (IPM) : The coordinated use of a combination of pest management methods to keep pest populations below economic injury levels.

Ion : One of the electrically charged particles into which certain molecules are separated or dissociated when in a solution of water. The molecules can be acids, bases and salts, e.g. paraquat dichloride separates into two negative chloride ions and a double-charged positive paraquat ion.

Larva(e) : The juvenile stage of an insect and related forms usually different in appearance to the adult and often a feeding phase causing significant damage, e.g. caterpillars.

LD 50 : Literally 'Lethal Dose 50%'. The amount of chemical (active constituent) expressed as milligrams per kilogram liveweight that will kill half a group of test animals. Usually administered orally (Acute oral LD_{50}) or to the skin (Acute dermal or percutaneous LD_{50}).

Lepidoptera : The order of insects that includes butterflies and moths. The larval stages are called caterpillars and cause the major portion of insect damage to growing crops.

Life cycle : The successive life stages of a plant or animal as they progress from birth to death.

Maximum residue limit (MRL) : The maximum legal amount of agricultural chemical residue expressed in mg of residue per kg of food commodity (1 mg/kg = 1 ppm) that is permitted to be present in marketed produce. MRLs are set in Australia by the APVMA and FSANZ authorities.

Metamorphosis : Significant charge of form passed through in a short time when a larval stage alters into a strikingly different adult shape. Butterfly and moth life cycles provide good examples of complete metamorphosis.

Microbial insecticides : Strictly these are insecticides based on bacterial action or toxins produced by bacteria. The expression sometimes is used to describe other living organisms as opposed to chemical agents, such as viruses and fungi used for pest control purposes.

Miticides : Agricultural chemicals that kill mites.

Mode of action : The manner in which an agricultural chemical controls pests. Invariably, this involves the disruption of some metabolic process in the target pest organism. For example, insect growth regulators prevent juvenile forms from reaching maturity and reproducing.

Molluscicides : Agricultural chemicals designed to kill slugs and snails (molluscs).

Monocotyledon (monocots) : A sub-class of plants that have a single first seed leaf. Most grasses and some members of the lily family fall within this category.

Monoculture : A farming system based almost entirely on a single crop.

Moulting : In caterpillars, the process of sloughing off the skin and acquiring a new and larger one as the body expands with feeding and growth. Each moult marks the end of an instar stage.

Mycelium : A mass of branching hyphae making the main vegetative body of a fungus.

Nematicides : Agricultural chemicals used to control nematodes (Class Nematoda), in particular the small soil-dwelling species that parasitise plants and are also known as eelworms.

Non-ionic : A molecule that does not form ions when placed in water. Nearly all wetting agents used in agriculture are non-ionic.

Non-polluting : Chemicals that break down quickly and are reabsorbed into natural chemical cycles without disruption to existing balances.

Ovicidal sprays (ovicides) : Agricultural chemical sprays intended to kill eggs (ova), usually of pest moths, butterfly and mites.

Pathogen : Disease-producing organism affecting a plant (or animal). It includes fungi, bacteria and viruses, but not higher organisms such as nematodes (Helminths).

Perennials : Plants with life cycles that persist for more than two years. Most flower regularly on an annual basis.

Phloem : The main transport system in vascular plants that carries food produced by photosynthesis throughout the plant. It consists of living cells, which means transported material must pass through a series of membranes as it moves.

Photosynthesis : The name given to the complex two-stage process used by chlorophyll bearing plants to use sunlight as an energy source to combine carbon dioxide and water to produce sugars and surplus oxygen.

Phylum : In animal classification, the word given to a group of diverse species which have a common body form. In plants the equivalent is a division.

Post-emergent herbicide : A herb tide applied after the crop to be treated or weed target has emerged after germination.

Pre-emergent herbicide : A herbicide applied to the soil and perhaps incorporated before the crop or target weeds have appeared above the soil surface. These herbicides persist until germination takes place.

Pre-harvest desiccation : Desiccant chemicals (e.g. diquat) applied to mature green crops to reduce moisture content and facilitate harvesting operations.

Protestant fungicide : A fungicide chemical designed to prevent infection when applied to plan: surfaces before the arrival and development of the pathogen.

Pupa(e) : The stage in an insect life cycle between larva and adult when activity is confined to the changes of body shape taking place within the pupa's outer covering.

Receptor site : In biochemistry, this is the physical site at which molecules meet and react in the course of life processes. The specific nature of some of these kites dictates which molecules can and which cannot proceed to react. Some agricultural chemicals depend for their action on receiving receptor sites and becoming involved in processes in a sinister way.

Resistance : The appearance of a high level of tolerance to an agricultural chemical in a pest species. This is likely to occur when the pest population has been subject to high selection pressure by repeated exposure to agricultural chemicals with a similar mode of action.

Rhizomes : A horizontal underground stem having buds at intervals that can produce shoots above ground and roots below.

Rodenticides : Agricultural chemicals used to control rats and mice.

Sanitation : Procedures designed to minimise disease spread, e.g. in nurseries by treating working surfaces with bactericides, fungicides etc., and the use of foot-baths to prevent entry into production areas of external disease agents on shoes.

Selective chemicals : Agricultural chemicals that have the ability to selectively target a pest without affecting the crop in which the pest is present.

Septa : Membrane walls that occur in some fungal mycelia dividing the tubular structure into segments.

Sequestering agent : In formulations, adjuvants designed to differentially combine with certain metallic elements, isolate them in soluble compounds and prevent the precipitation of solid particles, which could cause blockages in filters and nozzles.

Soft weeds : See Herbaceous weeds.

Soluble powder : An agricultural chemical formulation which is packaged as a powder and which dissolves completely in water to form a spray solution.

Solution : A liquid containing one or more compounds in a completely homogenous state. Usually refers to agricultural chemicals dissolved in and mixed with water-an aqueous solution.

Sorptive dusts : Dusts that attract and absorb moisture creating a desiccating atmosphere (in a silo).

Spores : A small usually one-celled body from which a new individual can develop. Produced by fungi, some plants such as ferns, protozoa and bacteria.

Spreader : In agricultural chemicals, a wetting agent added to a spray to improve target coverage.

Sticker : A material included in, or added to, an agricultural chemical formulation that increases the retention on surfaces to which the spray is applied.

Stolons : Sometimes called runners. Slender above ground horizontal stems that spread out from a plant.

Stomach poison : A poison that must be eaten to produce an effect. It will not kill by contact.

Stomata

(singular stoma) : Microscopic pores or holes in the epidermis (skin) of leaves and other plant parts through which the plant takes in carbon dioxide and gives out oxygen. Water loss in plants also occurs via the stomata.

Surfactant : Short for '**Surf**ace **Act**ive **Agent**'. This term is used to describe wetting agents, stickers and spreaders. They are usually nonionic if used with agricultural chemicals.

Suspension concentrate : See Flowable formulation.

Synergist : A chemical that increases the biological effect of another when the two are mixed.

Systemic fungicides : Fungicides with the ability to penetrate plants on which they are applied and move within those plants to attack invasive fungi.

Systemic insecticides : Insecticides with the ability to penetrate plants on to which they are applied and move with and become part of the plant's sap. They provide a lethal dose to sucking insect pests when they draw in that sap.

Taproot : The prominent main root of a plant's root system. Everyday examples of taproots include carrots and parsnips.

Transgenic crops : Crops that have had exotic genetic material engineered into their genetic code to express some desirable production trait. The best-known example is Monsanto's Bollgard cotton, which contains the code from Bacillus thuringiensis which produces a protein toxic to Helicoverpa spp.

Translocation : The term used to describe the movement of chemicals within growing plants.

Tuber : A modified fleshy and thickened underground root used for storage purposes. The potato is a common example of a tuber.

Ultra low volume (ULV) : A spraying technique that utilises equipment delivering spray volumes at 5 to 3 L/ha or less with droplet size below 150 microns.

Vascular system : The collective term used to describe the liquid transporting organs of plants, the phloem and the xylem. In animals, it is used to describe the vessels carrying the blood and lymphatic systems.

Vegetative reproduction : A non-sexual means of reproduction exhibited by some plants. Rhizomes, stolons and bulbs can all contribute towards this means of a plant spreading especially when an established perennial plant is cultivated and split up.

Water miscible : Another liquid able to mix completely with water to form a homogenous mixture.

Weed : Any plant growing where it is not wanted, including wild species, garden escapees and volunteer crop plants. The presence of weeds invariably creates a problem for human activities, e.g. by reducing the productivity or viability of the infested area, poisoning animals or contributing to environmental degradation.

Wettable powder : An agricultural chemical formulation designed to form a suspension when mixed with water to make up an agricultural chemical spray solution.

Wetter : See Surfactant.

Woody weeds : Weeds of a substantial erect nature often shrub or tree-like in growth habit. They are characterised by the formation of protective bark around their stems, which makes chemical penetration more difficult than in the case of soft weeds.

Xylem : The transport system in plants used for movement of water and dissolved and dissolved salts from the roots and throughout the plant. It consists of hollow tubes of dead thickened (lignified) tissue that give some strength to the stem structure. The flow in the xylem is upwards and outwards.